Rings with
Polynomial Identities

PURE AND APPLIED MATHEMATICS

A Series of Monographs and Textbooks

Rings with Polynomial Identities

Claudio Procesi

Universita di Pisa
Pisa, Italy

MARCEL DEKKER, INC.   New York   1973

3 3001 00572 9412

Alla mia Nanina

CONTENTS

PREFACE

The aim of the book is to provide a systematic exposition of a set of results obtained in the last twenty five years in a particular part of non commutative ring theory, i.e., the theory of rings with polynomial identities, briefly *PI*-rings.

A *PI*-ring is roughly a ring R for which there is a non 0 polynomial $f(x_1, \ldots, x_n)$ vanishing identically when computed in R. These rings appear in a natural way in algebra since a commutative ring is a *PI*-ring and given a *PI*-ring A any subring, quotient or finite matrix ring constructed from A is again a *PI*-ring. Therefore starting from commutative rings one can construct with the previous operations many rings which fall under the scope of our study (in fact one can show that the class of PI-rings is larger than the one previously described). This is only one reason to study *PI*-rings; the second lies in representation theory. In fact given a finite dimensional representation of a ring R in a matrix ring, say $(C)_n$, one sees that certain polynomials, evaluated in R, give elements vanishing under the representation and these polynomials depend only on n. For instance for $n = 1$ all commutators give rise to such elements. The interplay between polynomial identities and representations is even stricter when one considers irreducible representations.

Therefore the plan of the book is as follows. We begin with an analysis of the categorical properties of *PI*-rings and we familiarize with some formal constructions and manipulations. Next we study special types of *PI*-rings and develop the necessary structure theorems. The theory developed is then applied to study irreducible representations of rings. Then we study some special topics like the theory of finitely generated *PI*-algebras, the algebras of generic matrices, and some algrebraic geometric settings associated to these objects. Finally some problems of combinatorial nature associated with the Kurosh problem are studied and the intrinsic characterization of Azumaya algebras in terms of polynomial identities is presented. In the last chapter we present some recent work by E. Formanek on the center of *PI*-rings and applications to representation theory.

Some remarks are in order, as some theorems like the Posner theorem can be viewed as special cases of theorems on more general types of rings and we have avoided these more general settings to present the material in a more compact form. The results of Formanek are destined to influence a future presentation of the basic parts of the whole subject. There have been recently various results on the theory of rings with involutions with polynomial identity which would require a separate treatment.

Finally the very interesting work of A. Regev on the tensor product of *PI*-rings has been omitted since its exposition would take considerable space and is independent of the rest of the material presented.

The theory presented is far from having a final form and there are still many research problems to be worked out, some of which are probably quite hard and require completely new techniques. A list of questions is provided at the end with some comments.

Claudio Procesi

Pisa, Italy

Rings with
Polynomial Identities

Chapter I

POLYNOMIAL IDENTITIES IN ALGEBRAS

We want to develop, first of all, the categorical properties of rings with polynomial identities to give an idea of their place in algebra.

We will work generally in the category $_\Lambda\mathscr{A}$ of associative algebra over a commutative ring Λ with 1. We do not assume our algebras to have a 1 themselves, but it is clear that all the theory we develop can be developed in a parallel way for the category of algebras with 1.

One has to be especially careful with the use of symbols. One example is the concept of the free algebra in a set of variables x_i; this concept is valid for both categories. This free algebra will be denoted by $\Lambda\{x_i\}$ with the warning that it may indicate two different rings. In fact if we work in the category of algebras with 1 then $\Lambda\{x_i\}$ has a 1 and it is a graded algebra with Λ in degree 0; for the category of algebras without 1 it is the augmentation ideal of the same algebra.

If A, $B \in {}_\Lambda\mathscr{A}$, we indicate by $\mathscr{M}_\Lambda(A, B)$ or $\mathscr{M}(A, B)$ the set of algebra homomorphisms between A and B; again, in case we work with algebras with 1 the same symbol will mean homomorphisms preserving 1.

§1   Varieties of Algebras

The definitions and theorems that will be given are in accordance with the general theory of universal algebras [31], and they could be developed in general for algebraic categories. We shall study the case of associative rings, which are the main objects of interest in this book.

Definition 1.1   A nonempty class $\mathscr{C}$ of algebras is called a variety if it satisfies the following axioms:

(V$_1$)   If $C \in \mathscr{C}$ and $B \to C$ is injective, then $B \in \mathscr{C}$.
(V$_2$)   If $C \in \mathscr{C}$ and $C \to B$ is surjective, then $B \in \mathscr{C}$.
(V$_3$)   If $C_\alpha \in \mathscr{C}$, with $\alpha \in I$, is a family of algebras, then $\prod_{\alpha \in I} C_\alpha \in \mathscr{C}$.

Given a class $\mathscr{S}$ of algebras, they generate a variety $\bar{\mathscr{S}}$ constructed in the following way: $B \in \bar{\mathscr{S}}$ if and only if there exists: (1) a family of algebras $C_\alpha \in \mathscr{S}$; (2) an injection $C \to \prod C_\alpha$; (3) a surjection $C \to B$.

Proposition 1.2   The class $\bar{\mathscr{S}}$ is a variety containing $\mathscr{S}$. If $\mathscr{C}$ is a variety containing $\mathscr{S}$ then $\mathscr{C}$ contains $\bar{\mathscr{S}}$.

Proof   It is clear that $\bar{\mathscr{S}} \supseteq \mathscr{S}$ and if a variety $\mathscr{C} \supseteq \mathscr{S}$ then $\mathscr{C} \supseteq \bar{\mathscr{S}}$. Finally, to verify that $\bar{\mathscr{S}}$ is a variety is straightforward and we leave it to the reader.

Given an algebra C, we indicate by $\{C\}$ the variety that it generates.

Definition 1.3   (1)   If $\{C\} \supseteq \{B\}$ we say that B is a specialization of C.
(2)   If $\mathscr{V}$ is a variety and $\mathscr{V} = \{C\}$ we say that C is a generic algebra in $\mathscr{V}$.
(3)   If $\{C\} = \{B\}$ we say that C is equivalent to B.

Problem   Given a variety $\mathscr{V}$, can we find a generic algebra in $\mathscr{V}$? The answer is positive, as we will see in **2.11.**

Theorem 1.4   Given a variety $\mathscr{V}$, the canonical inclusion functor $i: \mathscr{V} \to {}_\Lambda\mathscr{A}$ has an adjoint (a retraction) $p: {}_\Lambda\mathscr{A} \to \mathscr{V}$.

Proof   Let R be an algebra. Consider the set $\mathscr{J} = \{I | I$ ideal of R, $R/I \in \mathscr{V}\}$. The set $\mathscr{J}$ is not empty since $R \in \mathscr{J}$. Furthermore, $\mathscr{J}$ is closed under arbitrary intersections since, if $I_\alpha \in \mathscr{J}$ is a family of ideals, then we have an injection $R/\bigcap I_\alpha \to \prod R/I_\alpha$ so that, by axioms (V_1), and (V_3), we obtain $R/\bigcap I_\alpha \in \mathscr{V}$. Therefore, $J_R = \bigcap_{I \in \mathscr{J}} I \in \mathscr{J}$. Then, by (V_3), and (V_2), we have $\mathscr{J} = \{I \mid I \text{ ideal of } R, I \supseteq J_R\}$. We set $p(R) = R/J_R$, and $p_R: R \to R/J_R$ the canonical projection. We have to verify that (i) we get a functor and (ii) we have an adjoint of the inclusion.

(i) If $\varphi: R \to S$ is a map of algebras, consider the composition

$$\psi = p_S \cdot \varphi, \psi: R \to S \to S/J_S.$$

We have an injection $R/\mathrm{Ker}\ \psi \to S/J_S$. Since $S/J_S \in \mathscr{V}$, by (V_1) we have $R/\mathrm{Ker}\ \psi \in \mathscr{V}$ so that $\mathrm{Ker}\ \psi \supseteq J_R$ and we can complete the diagram:

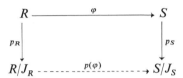

The verification that we have thus defined a functor is now trivial.

(ii) Let R be an algebra, $S \in \mathscr{V}$ and $\varphi: R \to S$. Then $J_S = 0$ and so we factor uniquely the map φ:

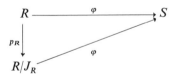

We have thus shown the universal property of p_R which implies (ii).

From the previous theorem it follows that a variety $\mathscr{V}$, considered itself as a category, has "free algebras." In fact, given a set S, we construct first the free associative algebra over S, $V = \Lambda\{x_s\}_{s \in S}$. Then the algebra $U = p(\Lambda\{x_s\})$ is a free algebra in the category $\mathscr{V}$ in the usual sense:

(1) $U \in \mathscr{V}$.
(2) We have a set theoretical map $s \to p_V(x_s) = \bar{x}_s$ the image of x_s in $V/J(V)$, $S \to U$.
(3) Given any other $R \in \mathscr{V}$, the set of maps $\mathscr{M}(U, R)$ is in a one-to-one

correspondence with the set R^S in the following way: $\varphi: U \to R$ corresponds to $(\varphi(\bar{x}_s))_{s \in S}$.

Examples

(a) The class of all commutative algebras is a variety. Given an algebra R, then J_R is the ideal generated by all commutators. The free algebras in this category are the usual polynomial algebras.

(b) The class of algebras with trivial product $xy = 0$, x, $y \in R$, is a variety. This category can be identified with the category of Λ modules. The free algebras are just the free Λ modules. The ideal $J_R = R^2$.

These examples are still rather trivial but we shall soon see the real scope of the theory.

§2   *T* Ideals, Polynomial Identities

Let $\mathscr{V}$ be a variety of algebras. Let R be any algebra, and J_R the minimal ideal of R such that $R/J_R \in \mathscr{V}$. The ideal J_R enjoys the following property.

Proposition 2.1   If $\varphi: R \to R$ is an endomorphism, then $\varphi(J_R) \subseteq J_R$.

Proof   This is just a special case of **1.4**(i).

Definition 2.2   An ideal $I \subseteq \Lambda\{x_s\}$ of the free algebra in the variables x_s is called a *T* ideal if, given any endomorphism $\varphi: \Lambda\{x_s\} \to \Lambda\{x_s\}$, we have $\varphi(I) \subseteq I$.

Corollary 2.3   Given a variety $\mathscr{V}$ and a free algebra $V = \Lambda\{x_s\}$, the ideal J_V is a *T* ideal.

We can describe the ideal J_V in a different way.

Proposition 2.4   The ideal J_V is the ideal of polynomials $f(x_s) \in \Lambda\{x_s\}$ with the property: given $R \in \mathscr{V}$ and elements $r_s \in R$, then $f(r_s) = 0$.

Proof   To give elements r_s is the same as to give a map $\varphi: \Lambda\{x_s\} \to R$ with $\varphi(x_s) = r_s$. From **1.4**(ii) we see that $\varphi(J_V) = 0$, so that if $f(x_s) \in J_V$ then $f(r_s) = 0$. Conversely, if $f(x_s) \in \Lambda\{x_s\}$ vanishes for any such φ, consider in particular the projection map $p: \Lambda\{x_s\} \to \Lambda\{x_s\}/J_V$, we have $p(f) = 0$ so that $f \in J_V$.

We have seen how a variety determines a T ideal; now we want to see the converse. Let $\Lambda\{x_s\}$ be a free algebra and $U \subseteq \Lambda\{x_s\}$ a set of polynomials.

Proposition 2.5
(1)   The class $\mathscr{C}$ of algebras R, such that for every $\varphi: \Lambda\{x_s\} \to R$ one has $\varphi(U) = 0$, is a variety.
(2)   The T ideal in $\Lambda\{x_s\}$ associated to $\mathscr{C}$ is the minimal T-ideal containing U.

Proof   (1)   We have to verify (V_1), (V_2), and (V_3).

(V_1) Let $C \in \mathscr{C}$ and $B \to C$ be an injection. If $\varphi: \Lambda\{x_s\} \to B$ is a map then the composition $\psi: \Lambda\{x_s\} \to B \to C$ is such that $\psi(U) = 0$. Since $B \to C$ is injective we must have $\varphi(U) = 0$.
(V_2) Let $C \in \mathscr{C}$ and $C \to B$ be surjective. If $\varphi: \Lambda\{x_s\} \to B$ is a map we can lift it to $\tilde{\varphi}: \Lambda\{x_s\} \to C$. Now $\tilde{\varphi}(U) = 0$, and therefore $\varphi(U) = 0$.
(V_3) If $C_\alpha \in \mathscr{C}$, $\alpha \in I$, consider the product $\prod C_\alpha$ and a map $\varphi: \Lambda\{x_s\} \to \prod C_\alpha$. For every projection $\pi_\alpha: \prod C_\alpha \to C_\alpha$ we have $\pi_\alpha \varphi(U) = 0$, therefore $\varphi(U) = 0$.
(2)   Let J be a T ideal containing U; consider the algebra $\Lambda\{x_s\}/J$; we have to show that $\Lambda\{x_s\}/J \in \mathscr{C}$. Now if $\varphi: \Lambda\{x_s\} \to \Lambda\{x_s\}/J$ is a map, we can lift φ to a map $\tilde{\varphi}: \Lambda\{x_s\} \to \Lambda\{x_s\}$. Since J is a T ideal and $U \subseteq J$, we have $\tilde{\varphi}(U) \subseteq J$ so that $\varphi(U) = 0$.

We want now to specialize the previous discussion to an algebra R and the variety $\{R\}$.

Definition 2.6   The set $I = \{f(x_s) \in \Lambda\{x_s\} | f(r_s) = 0$ for all $r_s \in R\}$ is called the set of polynomial identities of R.

Proposition 2.7   If $\mathscr{V} = \{R\}$ and J_V is the T ideal in $\Lambda\{x_s\}$ associated to $\mathscr{V}$, then $J_V = I$.

Proof   From **2.4** it is clear that $J_V \subseteq I$. By **2.5**, the set I determines a class of algebras $\mathscr{C}$, which is a variety for which the elements of I are polynomial identities. Now $R \in \mathscr{C}$ so that $\{R\} \subseteq \mathscr{C}$. Therefore it follows that $I \subseteq J_V$.

We want to sum up our results with a conclusive theorem but we still need a further proposition of a general nature.

Proposition 2.8   Every variety $\mathscr{V}$ is closed under inverse and direct limits.

Proof   Since $\mathscr{V}$ is closed under products it is clearly closed under inverse limits.

Let $\lambda_{\beta\alpha} \colon A_\alpha \to A_\beta$ be a directed set, let $A = \varprojlim A_\alpha$, and $f_\alpha \colon A_\alpha \to A$ be the canonical maps. We assume that all $A_\alpha \in \mathscr{V}$ and, since $\mathscr{V}$ is closed under surjections, we have $\bar{A}_\alpha = f_\alpha(A_\alpha) \in \mathscr{V}$. Therefore what we have to prove is that $A = \bigcup \bar{A}_\alpha \in \mathscr{V}$, where the $\bar{A}_\alpha$'s are an inductive family of subalgebras.

Let $V = \Lambda\{x_s\}$ be any free algebra and let J_V be the T ideal associated to $\mathscr{V}$ in V. Let I be the ideal of polynomial identities of A. We claim that $J_V \subseteq I$. Therefore let $f \in J_V$ and $r_s \in A$. We want to compute $f(r_s)$. It is clear that the value of $f(x_s)$ in $\{r_s\}$ depends only on a finite number of indices $s_1, \ldots, s_k$, those of the variables actually appearing in f. Since the $\bar{A}_\alpha$ are an inductive family we can find a suitable index α_0 such that $r_{s_1}, \ldots, r_{s_k} \in \bar{A}_{\alpha_0}$. Now $f(r_s)$ is really a polynomial computed in $\bar{A}_{\alpha_0}$ and since $\bar{A}_{\alpha_0} \in \mathscr{V}$ we have $f(r_s) = 0$. This shows that $J_V \subseteq I$.

This proof, in fact, shows something more, i.e., if $A = \bigcup \bar{A}_\alpha$, with $\bar{A}_\alpha$ an inductive family, then the ideal of polynomial identities of A is the intersection of the ideals of polynomial identities of the $\bar{A}_\alpha$'s.

Now, to finish the proof, assume we have taken so many variables that we can find a surjection $\pi \colon \Lambda\{x_s\} \to A$. Since J_V is an ideal of polynomial identities for A, we have $\pi(J_V) = 0$ so that the map factors through a $\tilde{\pi} \colon \Lambda\{x_s\}/J_V \to A$. Now $\Lambda\{x_s\}/J_V \in \mathscr{V}$ so that by (V_2), we obtain $A \in \mathscr{V}$.

Corollary 2.9   An algebra R is in a variety $\mathscr{V}$ if and only if every finitely generated subalgebra of R is in $\mathscr{V}$.

We now can prove the main theorem of this section.

Theorem 2.10 Let $V = \Lambda\{x_s\}$ be a free algebra in infinitely many variables.

(1) To every T ideal U of V corresponds the variety of algebras satisfying the polynomial identities in U.

(2) To every variety of algebras corresponds the T ideal of polynomial identities of this variety.

(3) In this way we establish a $1 - 1$ correspondence between the set of T ideals of V and the class of varieties of algebras.

Proof (1) and (2) are restatements of propositions already proved; the main point of the theorem is (3). We proceed, therefore, to prove (3). Let U be a T ideal, $\mathscr{V}$ the variety associated to U. By **2.5**(2) we know that the T ideal associated to $\mathscr{V}$ is again U, so one part of (3) was already proved. We have to show that if $\mathscr{V}$ is a variety and U the T ideal associated in V, then the class $\mathscr{C}$ of algebras defined by U is exactly $\mathscr{V}$.

Clearly $\mathscr{V} \subseteq \mathscr{C}$ by **2.4**. Let $R \in \mathscr{C}$; to show that $R \in \mathscr{V}$ we can use the criterion given by **2.9**. We can, therefore, assume directly that R is finitely generated. Let $R = \Lambda\{a_1, \ldots, a_t\}$ (this symbol will clearly indicate the algebra generated by the elements a_i's over Λ). Since we have an infinite set of variables, we can choose t elements $x_{s_1}, \ldots, x_{s_t}$ among the $\{x_s\}$ and we can prescribe a surjection $\varphi: \Lambda\{x_s\} \to R$ sending $x_{s_i} \to a_i$; φ is defined arbitrarily on the other variables. Since $R \in \mathscr{C}$ we have $\varphi(U) = 0$; thus we have a surjection $\Lambda\{x_s\}/U \to R$. Since $\Lambda\{x_s\}/U \in \mathscr{V}$, by (V_2) we have $R \in \mathscr{V}$.

Corollary 2.11 If $\mathscr{V}$ is a variety, $V = \Lambda\{x_s\}$ a free algebra in infinitely many variables, and J_V the T ideal associated to $\mathscr{V}$, then V/J_V is generic in $\mathscr{V}$ (i.e., $\{V/J_V\} = \mathscr{V}$).

Proof Clearly $\{V/J_V\} \subseteq \mathscr{V}$; on the other hand, these two varieties determine the same T ideal J_V in V so that $\mathscr{V} = \{V/J_V\}$.

Remark 2.12 Let $\mathscr{V}$ be a variety, $V = \Lambda\{x_s\}$ a free algebra, and J_V the T ideal associated to $\mathscr{V}$. Clearly, $\{V/J_V\} \subseteq \mathscr{V}$; but we need not necessarily have $\{V/J_V\} = \mathscr{V}$ if the set S of variables is finite. We shall see that in some specially interesting cases we have equality even if S has two elements (one element would imply that all algebras in the variety are commutative).

Exercises

(1) $\Lambda\{x_1, x_2\}$ is generic in the variety of all algebras.

(2) If F is an infinite field and $F\{x\}$ is the free algebra (without unit) in one variable, then the T ideals of $F\{x\}$ are just the ideals (x^n) for all n.

We finish with a technical proposition.

Proposition 2.13   Let $\varphi: \Lambda \to \Gamma$ be a map of commutative rings, then any Γ algebra can be considered as a Λ algebra.

(1) If A, B are Γ algebras and if A is a specialization of B in $_\Gamma\mathscr{A}$, then the same is true in $_\Lambda\mathscr{A}$.

(2) If $U \subseteq \Gamma\{x_s\}$ is a T ideal, the preimage of U in $\Lambda\{x_s\}$ is again a T ideal.

(3) If A is a Γ algebra, the T ideal of polynomial identities of A with coefficients in Λ is the preimage of the T ideal of polynomial identities of A with coefficients in Γ.

Proof   (1) This is clearly trivial because if C, D are Γ algebras and $\varphi: C \to D$ is a map, then C, D are Λ algebras and φ is a map of Λ algebras, so that the variety of Λ algebras generated by B contains the variety of Γ algebras generated by B.

(2) Consider the map $\varphi_*: \Lambda\{x_s\} \to \Gamma\{x_s\}$ induced by φ. Let $U \subseteq \Gamma\{x_s\}$ be a T ideal and consider $\varphi_*^{-1}(U)$. If $\psi: \Lambda\{x_s\} \to \Lambda\{x_s\}$ is an endomorphism, set $a_s = \varphi_*\psi(x_s)$. There is a unique endomorphism of Γ algebras $\bar{\psi}: \Gamma\{x_s\} \to \Gamma\{x_s\}$ for which $\bar{\psi}(x_s) = a_s$. Clearly the diagram

$$
\begin{array}{ccc}
\Lambda\{x_s\} & \xrightarrow{\psi} & \Lambda\{x_s\} \\
\varphi_* \downarrow & & \downarrow \varphi_* \\
\Gamma\{x_s\} & \xrightarrow{\bar{\psi}} & \Gamma\{x_s\}
\end{array}
$$

is commutative. Since U is a T ideal, $\bar{\psi}(U) \subseteq U$ so that

$$\varphi_*(\psi(\varphi_*^{-1}(U)) = \bar{\psi}\varphi_*(\varphi_*^{-1}(U)) = \bar{\psi}(U) \subseteq U,$$

hence $\psi(\varphi_*^{-1}(U)) \subseteq \varphi_*^{-1}(U)$.

(3) If A is a Γ algebra, and it is considered as a Λ algebra via the map

$\varphi: \Lambda \to \Gamma$, then clearly any map $\Lambda\{x_s\} \to A$ factors uniquely as $\Lambda\{x_s\} \to \Gamma\{x_s\} \to A$ so that $f \in \Lambda\{x_s\}$ vanishes on A if and only if $\varphi_*(f)$ is a polynomial identity of A.

§3   Special Types of Identities

We begin with an example. Let Λ be a commutative ring and I an ideal of Λ. Consider $I\{x_s\} \subseteq \Lambda\{x_s\}$. The ideal $I\{x_s\}$ is the ideal of $\Lambda\{x_s\}$ formed by all polynomials with coefficients in I. We have $\Lambda\{x_s\}/I\{x_s\} = \Lambda/I\{x_s\}$.

Proposition 3.1
(1)   The ideal $I\{x_s\}$ is a T ideal.
(2)   The variety corresponding to $I\{x_s\}$ is the variety of all algebras annihilated by I.
(3)   This variety can be identified with the category of all Λ/I algebras.

Proof   The proof is immediate and it is left to the reader.

It is quite clear from this example that, unless we are careful in our study, we need not find well-behaved algebras in a variety. In fact, the variety of all Λ/I algebras, if $I \neq \{0\}$, is a proper variety of the variety of all algebras. But it is also just the category of all algebras for another commutative ring. In our further study we want to carefully avoid such a situation. We set a definition.

Definition 3.2
(1)   A variety $\mathscr{V}$ of Λ algebras is called a proper variety if it does not contain the class of all Λ/I algebras for any ideal $I \neq \Lambda$.
(2)   A T ideal U of $\Lambda\{x_s\}$ is a proper T ideal if it is not contained in $I\{x_s\}$ for any ideal $I \neq \Lambda$.

Proposition 3.3   A variety $\mathscr{V}$ is proper if and only if the corresponding T ideal is proper (we assume the set of variables to be infinite).

Proof   It follows directly from **2.10.**

There is another way of giving condition (2) of **3.2.** Given any polynomial $f(x)$ in $\Lambda\{x_s\}$, we can consider the ideal generated by all coefficients of this polynomial; we indicate this ideal by $c(f)$ (the content of f).

Proposition 3.4   A T ideal U is a proper ideal if and only if there is an $f \in U$ with $c(f) = \Lambda$.

Proof   If there is an $f \in U$ with $c(f) = \Lambda$, it is clear that U is proper. Conversely, if U is proper then the coefficients of the elements of U must generate Λ. Let $f_1, \ldots, f_s \in U$ be such that their coefficients generate Λ; then if we choose high enough powers of a variable, say x_1, we have

$$f = f_1 x_1^{h_1} + f_2 x_1^{h_2} + \cdots + f_s x_1^{h_3} \in U$$

and no monomial of $f_i x_1^{h_i}$ is similar to a monomial of $f_j x_1^{h_j}$ if $i \neq j$. Therefore, $c(f) = \sum c(f_i) = \Lambda$.

The main objects of our study are defined now.

Definition 3.5

(1)   An algebra R will be called an algebra with polynomial identities, abbreviated a *PI* algebra, if the T ideal of polynomial identities of R is proper.

(2)   A polynomial identity $f(x_s)$ of R is called nontrivial if $c(f)R \neq 0$.

(3)   A polynomial identity $f(x_s)$ of R is called proper if $c(f)r = 0$, $r \in R$, implies $r = 0$.

One of the results that we shall prove is the following: If R satisfies a proper identity then it is a *PI* algebra. For the moment we prove an auxiliary result.

Proposition 3.6   If a polynomial $f(x_s)$ is a proper identity for an algebra R, then it is also a proper identity for the algebra $\Lambda\{x_s\}/U$, where U is the T ideal of the identities of R.

Proof The polynamial $f(x_s)$ is clearly an identity for $\Lambda\{x_s\}/U$ (see **2.7**). Let $g \in \Lambda\{x_s\}$ be such that $c(f)g \in U$; we wish to show that $g \in U$, i.e., that g is a polynomial identity of R. Now let $r_s \in R$; we have $c(f)g(r_s) = 0$, since $c(f)g \in U$. The hypothesis on f implies that $g(r_s) = 0$, therefore $g \in U$.

We set some further definitions.

Definition 3.7 A polynomial $f(x_s)$ is called multilinear if every variable appears with degree at most 1 in every monomial.

Definition 3.8
(1) Two monomials are called equivalent if they are made by the same variables (this is clearly an equivalence relation).
(2) A polynomial $f(x_s)$ is called uniform if all monomials appearing in $f(x_s)$ are equivalent.

Given any polynomial $f(x_s)$ we can always group together equivalent monomials and write, in a unique way, $f(x_s) = \sum f_i(x_s)$ where $f_i(x_s)$ is uniform.

Proposition 3.9 Let U be a T ideal, $f(x_s) \in U$ and $f(x_s) = \sum_{i=1}^{k} f_i(x_s)$ be the canonical decomposition in uniform polynomials. Then $f_i(x_s) \in U$ for every i.

Proof Let us specify the variables actually appearing in $f(x)$ as $x_1, \ldots, x_h$. We induce on k. If $k = 1$ the polynomial is uniform. Otherwise there is at least one variable, say x_1, which appears in some of the $f_i(x)$, say $i = 1, \ldots, s$, and does not appear in the others, $i = s + 1, \ldots, k$. Set $x_1 = 0$, since U is a T ideal $f(0, x_2, \ldots, x_h) \in U$. Now $f(0, x_2, \ldots, x_h)$ $= \sum_{i=s+1}^{k} f_i(x_s)$. Therefore

$$\sum_{i=1}^{s} f_i(x) = f(x_1, \ldots, x_h) - f(0, x_2, \ldots, x_h)$$

and $\sum_{i=s+1}^{k} f_i(x)$ belong to U. By induction their uniform components are in U and the proposition is proved.

Corollary 3.10   If R satisfies a nontrivial identity then it satisfies a nontrivial uniform identity.

We now turn our attention to multilinear identities. We assume the number of variables to be infinite.

Theorem 3.11   If U is a nonzero T ideal then we can find a nonzero multilinear polynomial $g \in U$.

Proof   Let $p(x) \in U$, $p(x) \neq 0$. Let $x_1, \ldots, x_h$ be the variables appearing in p, $d_i(p) = $ degree of x_i in p; $L = L_p = \max d_i(p)$, and $s_p = $ number of i's such that $d_i(p) = L$. If $L = 1$ then $p(x)$ is multilinear; otherwise we make a double induction on the two numbers L, s. Say that the first variables $x_1, \ldots, x_s$ appear with degree L. Choose a new variable x_{h+1} and consider the polynomial

$$q(x_1, \ldots, x_h, x_{h+1}) = p(x_1 + x_{h+1}, x_2, \ldots, x_h)$$
$$- p(x_1, \ldots, x_h) - p(x_{h+1}, x_2, \ldots, x_h).$$

We claim that (i) $L_q \leqslant L_p$, (ii) $s_q \leqslant s_p - 1$, (iii) $q(x) \neq 0$. This will prove the theorem by induction.

(i) This is clear; in fact, $d_i(q) \leqslant d_i(p)$ for $i = 1, \ldots, h$, and $d_{h+1}(q) \leqslant L$.

(ii) It is enough to prove that $d_1(q)$, $d_{h+1}(q) < L$. Now the total degree of x_1 and x_{h+1} in composite in q is $\leqslant L$; to get the actual monomials in which x_1 or x_{h+1} appear with degree L we set x_1 or x_{h+1} equal to 0. Now

$$q(0, x_2, \ldots, x_{h+1}) = p(x_{h+1}, \ldots, x_h) - p(0, x_2, \ldots, x_h)$$
$$- p(x_{h+1}, \ldots, x_h) = -p(0, x_2, \ldots, x_h).$$

This polynomial does not contain x_{h+1}; therefore the total degree L is never attributed entirely to x_{h+1} in $q(x)$; the result is similar for x_1.

(iii) Consider a monomial $M(x) = x_{i_1} x_{i_2} \ldots x_{i_n}$ appearing in $p(x)$ with x_1 of degree L. The polynomial $M(x_1 + x_{h+1}, x_2, \ldots, x_h)$ is a sum of monomials $M_j(x_1, \ldots, x_h, x_{h+1})$ for each of which we have

$$M_j(x_1, \ldots, x_h, x_1) = M(x);$$

therefore, in the expansion of $p(x_1 + x_{h+1}, \ldots, x_h)$, terms M_j and M_i' are always different monomials for different monomials M and M'. Therefore, $p(x_1 + x_{h+1}, \ldots, x_h)$ is nonzero and contains monomials in both x_1 and x_{h+1} with nonzero coefficients; clearly such monomials continue to appear in $q(x)$.

Corollary 3.12 If R satisfies a nontrivial identity of degree d then R satisfies a nontrivial multilinear identity of degree $\leq d$.

Proof Let $p(x)$ be a nontrivial identity for R; write

$$p(x) = q(x) + s(x)$$

where all coefficients of $s(x)$ annihilate R and no coefficient of $q(x)$ annihilates R. The polynomial $q(x)$ is still an identity of R. If we multilinearize $q(x)$ according to the procedure of **3.11**, part (iii) of the proof whows that some coefficients of $q(x)$ appear in the first result of the multilinearization procedure for $q(x)$, and so this first polynomial is a nontrivial identity for R. We continue in the same way by induction and prove the corollary.

Proposition 3.13 The main property of multilinear identities is the following.

(1) Let $f(x) \in \Lambda\{x_s\}$ be a multilinear polynomial, R a Γ algebra, and $\varphi: \Lambda \to \Gamma$ a map of commutative rings. Let $S \subseteq R$ be a subset, spanning R as a Γ module. If $f(x)$ vanishes whenever computed on the elements of S, then $f(x)$ is a polynomial identity of R.

(2) If $f(x)$ is a multilinear identity of a Λ algebra T, and $\varphi: \Lambda \to \Gamma$ a map of commutative rings, then $f(x)$ is also an identity of $T \otimes_\Lambda \Gamma$.

Proof (1) Let $f(x)$ depend on the variables $x_1, \ldots, x_n$. Let $a_i = \sum_{j=1}^{s} r_{ij}\alpha_{ij}$, $i = 1, \ldots, n$, and let $r_{ij} \in S$ and $\alpha_{ij} \in \Gamma$ be n elements of $R = S\Gamma$. We have to prove that

$$f(a_1, \ldots, a_n) = 0. \text{ Now } f(a_1, \ldots, a_n)$$
$$= \sum \alpha_{1j_1}\alpha_{2j_2} \ldots \alpha_{nj_n} f(r_{1j_1}, r_{2j_2}, \ldots, r_{nj_n}) = 0$$

since $f(x)$ is multilinear and it vanishes on S.

(2)   It is a special case of (1), taking $R = T \otimes_\Lambda \Gamma$ and $S = T$.

We want to study a little more the identities which behave like the multilinear ones.

Definition 3.14   A polynomial $f(x)$ is called a stable identity for the algebra R if, for every map $\varphi \colon \Lambda \to \Gamma$ of commutative rings, $f(x)$ is an identity of $R \otimes_\Lambda \Gamma$.

Clearly the stable identities of R are a T ideal, being the common identities of a certain class of algebras; we determine now a property of stable identities.

Proposition 3.15
(1)   A polynomial $f(x)$ is a stable identity for the algebra R if and only if all its totally homogeneous parts are stable identities for R.
(2)   If $f(x)$ is a stable identity of R and $R \to S$ is a surjective map, then $f(x)$ is a stable identity of S.

Proof   (1)   Let $x_1, \ldots, x_n$ be the variables appearing in $f(x)$; for every n-tuple $T = (t_1, \ldots, t_n)$ of natural numbers, we consider $f_T(x)$ equal to the sum of all monomials of $f(x)$ with x_i of degree t_i. We have $f(x) = \sum_T f_T(x)$ and we have to prove that $f(x)$ is a stable identity for R if and only if $f_T(x)$ is a stable identity for R for all T. If for all T the $f_T(x)$ is a stable identity of R, clearly their sum is again a stable identity.
Conversely, let $\Gamma = \Lambda[\lambda_1, \ldots, \lambda_n]$ be the polynomial algebra in n variables. If $f(x)$ is a stable identity of R, then $f(\lambda_1 r_1, \lambda_2 r_2, \ldots, \lambda_n r_n) = 0$ for all $r_1, \ldots, r_n \in R$. Now

$$f(\lambda_1 r_1, \lambda_2 r_2, \ldots, \lambda_n r_n) = \sum \lambda^T f_T(r_1, \ldots, r_n)$$

so that $f_T(r_1, \ldots, r_n) = 0$ for all r_i's $\in R$. Therefore $f_T(x)$ is an identity of R. Now for any map $\Lambda \to \Pi$ it is clear that $f(x)$ is still a stable identity for $R \otimes_\Lambda \Pi$ so that $f_T(x)$ is an identity for $R \otimes_\Lambda \Pi$. This proves that all $f_T(x)$ are stable identities of R.
(2)   This follows from the last remarks made in the proof of (1): if $f_T(x)$ is an identity of R for all T's, it is also an identity of S for all T's and thus a stable identity of S.

A totally homogeneous polynomial need not be a stable identity.

Example Take the polynomial ring (without 1) $Z/(2)$ $[x, y]$, modulo the ideal generated by the elements z^3, $z \in Z/(2)[x, y]$. We have an algebra with a basis x, y, x^2, y^2, xy, x^2y and with $yx^2 = y^2x$. The identity z^3 is not stable since for a field $F \supseteq Z/(2)$ we have $(\alpha x + y)^3 = (\alpha^2 + \alpha)x^2y \neq 0$ as soon as $\alpha \notin Z/(2)$.

Remark Given a polynomial $f(x_1, \ldots, x_s)$, we add

$$m = k_1 + k_2 + \cdots + k_s$$

commutative indeterminates λ_{i,j_i}, $i = 1, \ldots, s$, $j_i = 1, \ldots, k_i$, to Λ and consider new variables y_{i,j_i} in the free algebra. Then

$$f(\sum \lambda_{1,j_1} y_{1,j_1}, \sum \lambda_{2,j_2} y_{2,j_2}, \ldots) = \sum \lambda^I f_I(y),$$

where I runs over all choices of m natural numbers α_{i,j_i} and $\lambda^I = \prod \lambda_{i,j_i}^{\alpha_{i,j_i}}$. Then it is clear that if $f(x)$ is a stable identity, $f^I(y)$ is a polynomial identity on R for all such choices; the converse is also clearly true.

We can make the situation a little clearer if we work over an infinite field. Let F be an infinite field and $F\{x_s\}$ the free algebra in the variables x_s.

Theorem 3.16
   (1)   If R is an F algebra, every polynomial identity of R is stable.
   (2)   Every T ideal I is homogeneous in all variables x_s.

Proof Clearly (1) $\Rightarrow$ (2) (**3.15**). We prove (1). Let I be the T ideal of identities of R and let $f(x_1, \ldots, x_n) \in I$. Let $H \supseteq F$ be a commutative ring, consider

$$a_1 = \sum \alpha_{i,1} r_{i,1}, \qquad a_2 = \sum \alpha_{i,2} r_{i,2}, \quad \ldots,$$

$$a_n = \sum \alpha_{i,n} r_{i,n} \in H \otimes_F R, \qquad \alpha_{i,j} \in H, r_{i,j} \in R$$

(we omit the $\otimes$ sign). Then $f(a_1, \ldots, a_n) = \sum \alpha^J f_J(r_{i,j})$, where α^J denotes a monomial in the elements $\alpha_{i,j}$. We can consider this a function of the $\alpha_{i,j}$'s fixing the $r_{i,j}$'s. If $\alpha_{i,j} \in F$, we have $f(a_1, \ldots, a_n) = 0$. Now F is an infinite field and, therefore, we must have $f_J(r_{i,j}) = 0$, and finally $\sum \alpha' f_J(r_{i,j}) = 0$ even if $\alpha_{i,j} \in H$, so that $f(x)$ is a stable identity.

Among multilinear identities there is a particular one which plays a fundamental role in the theory.

Definition 3.17   The polynomial (in $Z\{x_1, \ldots, x_n\}$),

$$S_n(x_1, \ldots, x_n) = \sum_{\sigma \in \mathscr{S}_n} \varepsilon(\sigma) x_{\sigma(1)} x_{\sigma(2)} \cdots x_{\sigma(n)},$$

where $\mathscr{S}_n$ is the symmetric group on n elements and $\varepsilon(\sigma)$ is the signature of σ (± 1 according that σ is even or odd), is called the standard polynomial of degree n.

Remark   We have $S_2(x_1, x_2) = x_1 x_2 - x_2 x_1$. Therefore an algebra satisfies S_2 if and only if it is commutative.

We develop a few properties of S_n. Let us write $S_n(X)$, $X = (x_1, \ldots, x_n)$.

Proposition 3.18
   (1) $S_n(x_{\sigma(1)}, x_{\sigma(2)}, \ldots, x_{\sigma(n)}) = \varepsilon(\sigma) S_n(x_1, x_2, \ldots, x_n)$.
   (2) $S_n(x_1, \ldots, x_n) = \sum (-1)^{i+1} x_i S_{n-1}(x_1, \ldots, \check{x}_i, \ldots, x_n)$.
   (3) $S_n(X) = \sum_{Y \subseteq X} \varepsilon(Y) S_h(Y) S_{n-h}(X - Y)$;
$\varepsilon(Y)$ is the signature of the permutation $(Y, X - Y)$.
   (4) $S_n(x_1, \ldots, x_n) = \sum_{i=3}^{n} (-1)^{i+1} S_{n-2}(x_1 x_2 x_i, x_3, \ldots, \check{x}_i, \ldots, x_n)$
$+ S_{n-2}(x_3, \ldots, x_n) x_1 x_2 + \sum_{\sigma \in T} \varepsilon(\sigma) x_{\sigma(1)} \cdots x_{\sigma(n)}$,
where T is the set of permutation's for which x_1 and x_2 are not consecutive.

Proof   The proof is a simple exercise on permutations; we note that (2) is a special case of (3).

Corollary 3.19   If we substitute in $S_n(x_1, x_2, \ldots, x_n)$ for x_i and x_j the variable y, $i \neq j$, we obtain 0.

Proof   Consider the transposition σ between x_i and x_j. Thus

$$S_n(x_{\sigma(1)}, \ldots, x_{\sigma(n)}) = -S_n(x_1, \ldots, x_n).$$

If we substitute for both members x_i and x_j the variable y, we conclude that the resulting value is equal to its opposite. Since the resulting polynomial has integer coefficients, it must be 0.

The first theorem from which we see the importance of standard identities is the following.

Theorem 3.20   Let R be an algebra over a commutative ring Λ. Assume that R is a Λ module of finite type, generated by n elements. Then R satisfies $S_{n+1}(x_1, \ldots, x_{n+1})$.

Proof   Let $r_1, \ldots, r_n \in R$ be linear generators of R over Λ, and

$$s_i = \sum_{j=1}^{n} \lambda_{ij} r_j \in R, i = 1, \ldots, n + 1.$$

Thus

$$S_{n+1}(s_1, \ldots, s_{n+1}) = \sum \lambda_{1,j_1} \lambda_{2,j_2} \cdots \lambda_{n+1,j_{n+1}} S_{n+1}(r_{j_1}, r_{j_2}, \ldots, r_{j_{n+1}}),$$

so that for each choice of the indices $j_1, \ldots, j_{n+1}$ among the numbers $1, \ldots, n$ there must always be a coincidence; therefore, by **3.19** we have $S_{n+1}(r_{j_1}, \ldots, r_{j_{n+1}}) = 0$ and so S_{n+1} is an identity of R.

Even if standard identities are of fundamental importance, there are still *PI-* algebras which do not satisfy any standard identity.

Example   Let $R = \wedge V$ be the exterior algebra on an infinite dimensional vector space V over a field F. We claim that R satisfies the identity $[x, [y, z]] = 0$. In fact, this polynomial is multilinear; therefore, it is sufficient to check that it vanishes on the elements of a basis of $\wedge V$. Let $a = v_1 \wedge \cdots \wedge v_k, b = w_1 \wedge \cdots \wedge w_s$. We have $[a, b] = 0$ if one of the two elements is of even degree. Therefore computing $[c, [a, b]]$, we see that either $[a, b] = 0$ or $[a, b]$ has even degree and therefore $[c, [a, b]] = 0$. Let us now show that if F has characteristic 0, then R does not satisfy any standard identity. Let $v_1, \ldots, v_n$ be a linearly independent family of vectors of V. Then

$$S_n(v_1, \ldots, v_n) = n!v_1 \wedge \cdots \wedge v_n \neq 0.$$

As a corollary, it is clear that $\wedge V$ cannot be embedded in any algebra which is a finite module over a commutative ring.

Another interesting instance in which we have polynomial identities is the following.

Definition 3.21   (1)   An algebra R over Λ is called integral over Λ if every element $r \in R$ satisfies a polynomial $r^n + \alpha_1 r^{n-1} + \cdots + \alpha_n = 0$ with $\alpha_i \in \Lambda$.

(2)   An algebra R is called integral of bounded degree N if every element satisfies a polynomial as in (1) of degree $n \leqslant N$.

Theorem 3.22   If R is an algebra of bounded degree N, then R satisfies the polynomial $S_{N+1}(y^N x, y^{N-1} x, \ldots, x)$.

Proof   Let $a, b \in R$ so that a satisfies a polynomial of degree N:

$$a^N + \alpha_1 a^{N-1} + \cdots + \alpha_N = 0, \alpha_i \in \Lambda.$$

Therefore

$$
\begin{aligned}
0 &= S_{N+1}(0, a^{N-1}b, a^{N-2}b, \ldots, b) \\
&= S_{N+1}(a^N b + \alpha_1 a^{N-1} b + \cdots + \alpha_N b, a^{N-1}b, \ldots, b) \\
&= S_{N+1}(a^N b, a^{N-1}b, \ldots, b) + \sum \alpha_i S_{N+1}(a^{N-i}b, a^{N-1}b, \ldots, b).
\end{aligned}
$$

Now all the terms $S_{N+1}(a^{N-i}b, a^{N-1}b, \ldots, b)$ are zero, since two entries are equal (**3.19**), so the theorem is proved.

§4   Equivalence and Specializations

We want to give some criteria for two algebras to satisfy the same identities. We will use the terminology given in **1.3**.

Proposition 4.1   Let R, S, and T be Λ algebras. Assume that S is a specialization of R, and T is Λ flat. Then $S \otimes_\Lambda T$ is a specialization of $R \otimes_\Lambda T$.

Proof   To say that S is a specialization of R means that we have a subalgebra U of a power R^I of R and a surjection $\pi: U \to S$. It follows that $U \otimes T \subseteq (R^I) \otimes T \subseteq (R \otimes T)^I$. The first inclusion is assured by the flatness of T. Then $\pi \otimes 1: U \otimes T \to S \otimes T$ is a surjection.

Corollary 4.2   Let R and S be two equivalent algebras. Then $(R)_n$ and $(S)_n$ (the algebras of $n \times n$ matrices) are equivalent.

Proof   We have $(R)_n \simeq R \otimes_\Lambda (\Lambda)_n$, and $(S)_n \simeq S \otimes_\Lambda (\Lambda)_n$, and $(\Lambda)_n$ is Λ flat.

We will use this corollary in particular when R is generic in the variety of commutative Λ algebras.

Definition 4.3   The T ideal associated to $(R)_n$, where R is a generic commutative algebra, is called the ideal of identities of $n \times n$ matrices with coefficients in Λ.

A particular case is when $\Lambda = Z/(p)$, where p is a prime number. In this case we can choose for R any infinite field of characteristic p and we can talk about the identities of $n \times n$ matrices of characteristic p. When $\Lambda = Z$ we can take $R = Z$ and we will talk simply about identities of $n \times n$ matrices.

Problem   If $Z\{x_s\}$ is a free algebra, I the ideal of identities of $n \times n$ matrices, and I_p the ideal of identities of $n \times n$ matrices in characteristic p, is $I_p = I + (p)\{x_s\}/(p)\{x_s\}$?

Proposition 4.4   Let B_1, B_2, and C be commutative Λ algebras and $\varphi_1: B_1 \to C$ and $\varphi_2: B_2 \to C$ be two maps. Let R_1 and R_2 be, respectively, B_1 and B_2 algebras. We assume:

(1) $R_1 \otimes_{\varphi_1} C$ is equivalent to $R_2 \otimes_{\varphi_2} C$ as C algebras;
(2) B_1 and C are equivalent as B_1 algebras;

(3) B_2 and C are equivalent as B_2 algebras;
(4) R_1 is B_1 flat, R_2 is B_2 flat.

Conclusion: The algebras R_1 and R_2 are equivalent as Λ algebras.

Proof   The algebra $R_1 = R_1 \otimes_{B_1} B_1$ is equivalent to $R_1 \otimes_{B_1} C$ as B_1 algebras since R_1 is B_1 flat and B_1 is equivalent to C **(4.1)**. Similarly, R_2 is equivalent to $R_2 \otimes_{B_2} C$ as B_2 algebras. Since $R_1 \otimes_{B_1} C$ is equivalent to $R_2 \otimes_{B_2} C$ as C algebras, we see **(2.13)** that R_1 and R_2 are Λ-equivalent.

Corollary 4.5   Let S_1 and S_2 be two simple algebras with centers Z_1 and Z_2; assume $\dim_{Z_1} S_1 = \dim_{Z_2} S_2 = n^2$. Assume further that Z_1, and Z_2 are infinite fields containing a fixed field K. Then S_1 and S_2 are equivalent over K.

Proof   We can embed Z_1 and Z_2 in a fixed algebraically closed field Ω. We have $S_1 \otimes_{Z_1} \Omega \simeq \Omega_n \simeq S_2 \otimes_{Z_2} \Omega$ so that **4.4** applies. All flatness hypotheses are satisfied since all the commutative algebras in question are fields; the equivalence of Z_i ($i = 1, 2$) with Ω is assured by the assumption that Z_i is an infinite field.

Finally, we give a criterion to determine with a finite procedure which polynomials are identities for a certain class of algebras.

Let R and S be Γ algebras. Assume that S is a free Γ module with basis $u_1, \ldots, u_m$. Let $U = \Gamma\{\bar{x}_{t,i}\} = \Gamma\{x_{t,i}\}/J$ be the free algebra in the variety generated by R, in the variables $x_{t,i}$, where $t \in T$, $i = 1, \ldots, m$ (with T a given set). Consider the algebra $U \otimes_\Gamma S$ and in this algebra the elements $\xi_t = \sum_{i=1}^{m} \bar{x}_{t,i} \otimes u_i$ (generic elements according to an ancient terminology). Let Λ be a commutative ring and $\Lambda \to \Gamma$ a map. We wish to determine the identities of $R \otimes_\Gamma S$ as a Λ algebra.

Theorem 4.6   The kernel of the map

$$\psi: \Lambda\{x_t\} \to U \otimes S, \text{ where } \psi(x_t) = \xi_t,$$

is the ideal of polynomial identities of $R \otimes_\Gamma S$. Therefore $\psi(\Lambda\{x_t\}) = \Lambda\{\xi_t\}$ is isomorphic to the free algebra in the variables x_t in the variety generated by $R \otimes_\Gamma S$ (as Λ algebra).

Proof   Given a homomorphism $\varphi: U = \Gamma\{\bar{x}_{s,i}\} \to B$ we obtain a morphism $\Lambda\{x_t\} \xrightarrow{\psi} U \otimes_\Gamma S \xrightarrow{\varphi \otimes 1} B \otimes_\Gamma S$. We claim that, in this fashion, we have a $1-1$ correspondence $(\varphi \leftrightarrow (\varphi \otimes 1) \cdot \psi)$ between Γ maps $\Gamma\{\bar{x}_{t,i}\} \to B$ and Λ maps $\Lambda\{x_t\} \to R \otimes_\Gamma S$. In fact, given a homomorphism $\gamma: \Lambda\{x_t\} \to R \otimes_\Gamma S$ we obtain $\gamma(x_t) = \sum_i r_{t,i} \otimes u_i$, where $r_{t,i} \in R$. Constructing $\tau: \Gamma\{\bar{x}_{t,i}\} \to B$ by the law $\tau(\bar{x}_{t,i}) = r_{t,i}$, we see that $\gamma = (\tau \otimes 1) \cdot \psi$. It is clear that in this way we have the inverse correspondence to $\varphi \to (\varphi \otimes 1) \cdot \psi$.

Now the assertion of the theorem follows since, if $g(x_t)$ is in the kernel of ψ, then it is in the kernel of every map $\Lambda\{x_t\} \to R \otimes_\Gamma S$. Conversely if $\psi(g(x_t)) = \sum g_i \otimes u_i \neq 0$, we must have $g_i \neq 0$ for some i. Therefore, since U is the free algebra in the category generated by B, there exists a $\varphi: U \to B$ for which $\varphi(g_i) \neq 0$. It follows that

$$(\varphi \otimes 1) \circ \psi(g(x)) = \sum \varphi(g_i) \otimes u_i \neq 0.$$

Apply this theorem in the special case of matrix algebras. Let $S = (\Gamma)_n$, where Γ is a generic commutative Λ algebra, $R = \Gamma$. For the basis of S, take $e_{i,j}$, $i, j = 1, \ldots, n$. Then $U = \Gamma[x_{t,i,j}]$ is the polynomial algebra in the variables $x_{t,i,j}$, with $t \in T$ and $i, j = 1, \ldots, n$. We have $U \otimes_\Gamma (\Gamma)_n = (U)_n$ and $\xi_t = \sum x_{t,i,j} e_{ij}$ is the matrix with entries $x_{t,i,j}$. We call the ξ_t's generic matrices. From this theorem it follows that the algebra $\Lambda\{\xi_t\} \subseteq (U)_n$ is isomorphic to the free algebra in the variety generated by $n \times n$ matrices over commutative algebras.

Finally it is useful to have some hold on generic commutative algebras.

Proposition 4.7   (1)   The ring $\Lambda[x]$ is a generic commutative Λ algebra.
   (2)   If Λ is a domain and $\Gamma \supseteq \Lambda$ is an infinite domain, then Γ is a generic commutative algebra.

Proof   (1)   We have to prove that if $f(x_1, \ldots, x_n) \in \Lambda[x_1, \ldots, x_n]$ is a nonzero commutative polynomial, we can evaluate f in $\Lambda[x]$ so as to make it nonzero. Let $x_i = x^{m_i}$; if

$$f(x_1, \ldots, x_n) = \sum \lambda_{i_1 i_2 \ldots i_k} x_1^{i_1} x_2^{i_2} \ldots x_n^{i_n}$$

then

$$f(x^{m_1}, x^{m_2}, \ldots, x^{m_n}) = \sum \lambda_{i_1 i_2 \ldots i_n} x^{\sum_{j=1}^{n} m_j i_j}$$

It is clear that one can choose the m_i's in such a way that all the numbers $\sum_{j=1}^{n} m_j i_j$ are distinct; therefore, $f(x^{m_1}, \ldots, x^{m_n}) \neq 0$.

(2) This is a well-known elementary fact. If $0 \neq f(x_1, \ldots, x_n) \in \Gamma[x_1, \ldots, x_n]$, with Γ an infinite domain, we can evaluate $f(x)$ in Γ so as to make it nonzero.

§5 The Identities of Matrix Algebras

Let Λ be a commutative ring with 1.

Theorem 5.1   The algebra $(\Lambda)_m$ does not satisfy any polynomial identity of degree less than $2m$.

Proof   If $f(x)$ is a nontrivial polynomial identity of degree $d \leq 2m$, we can deduce from this a multilinear one $g(x)$ which is also uniform (**3.9, 3.11**). Let d be the degree of $g(x)$ with $x_1, \ldots, x_d$ the variables appearing in $g(x)$. We can assume that, among the monomials appearing in $g(x)$, we have a monomial $\alpha x_1 x_2, \ldots, x_d$, with $\alpha \neq 0$. Substitute for $x_1, \ldots, x_d$ the elements $e_{11}, e_{12}, e_{22}, e_{23}, \ldots, e_{k\,k+1}$ if $d = 2k$ is even, or $e_{11}, \ldots, e_{k\,k}$ if $d = 2k - 1$ is odd. It is clear that the only product of these elements which in not zero is $e_{11}e_{12}e_{22} \cdots e_{k\,k+1} = e_{1\,k+1}$ (respectively, $e_{11}e_{22} \cdots e_{k\,k} = e_{1k}$). Therefore,

$$g(e_{11}, \ldots,) = \alpha e_{1\,k+1} \quad (\text{or } \alpha\, e_{1k}) \neq 0$$

is a contradiction.

The main theorem that we know about the identities of matrix algebras is the following.

Theorem 5.2 (Amitsur-Levitzki) $(\Lambda)_m$ satisfies $S_{2m}(x_1, \ldots, x_{2m})$.

We precede this theorem with a result on graph theory following R. Swan.

§6   A Theorem on Graphs

Definition 6.1

(1) An oriented graph Γ consists of two sets V and E whose elements are called respectively, vertices and edges, and two maps ε_1, $\varepsilon_2 : E \to V$.

(2) If $s \in E$, then $\varepsilon_1(s)$ is called the initial point of s and $\varepsilon_2(s)$ the terminal point.

(3) Given two vertices a, $b \in V$, an oriented path from a to b is a sequence $s_1, \ldots, s_t$ of edges such that $\varepsilon_1(s_1) = a$, $\varepsilon_2(s_t) = b$, and $\varepsilon_2(s_i) = \varepsilon_1(s_{i+1})$, with $i = 1, \ldots, t - 1$.

The relation "a can be connected to b by an oriented path" generates an equivalence relation in V.

If $(V_i)_{i \in I}$ are the equivalence classes for this relation, one easily sees that $\varepsilon_1^{-1}(V_i) = \varepsilon_2^{-1}(V_i)$ and, setting $E_i = \varepsilon_1^{-1}(V_i)$, we have that E is the disjoint union of the E_i's. Therefore, one sees that Γ can be considered as the disjoint union of the graphs (V_i, E_i). If there is only one class we say that Γ is connected.

Assume now that Γ is finite, i.e., that V and E are finite sets. Let $v = |V|$, $e = |E|$ be the respective numbers of elements.

Definition 6.2   If Γ is a finite graph, a unicursal path in Γ is a path $s_1, \ldots, s_t$ such that each edge appears one and only one time in the sequence (i.e., it is a bijective map $(1, \ldots, t) \to E$ which is also a path).

Choose an ordering of E; a unicursal path can then be considered as a particular permutation of the elements in E, and it makes sense to ask whether it is odd or even. Fixing further two vertices a, b we can consider the numbers n_1 and n_2, respectively, of even and odd unicursal paths starting in a and ending in b. The main result that we wish to prove is the following.

Theorem 6.3   If $e \geqslant 2v$ we have $n_1 = n_2$.

Proof   First we remark that $e \geqslant 2v$ is really necessary, for the graph

with v vertices and $2v - 1$ edges has only one unicursal path from a to a.

We prove the theorem, first making some reductions and then proceeding by induction.

(i) If $e > 2v$, let $k = e - 2v$; we modify the graph by attaching to a a string of k new edges

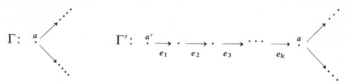

The new graph Γ' has $v' = v + k$ vertices, $e' = e + k$ edges, and there is a 1–1 correspondence between unicursal paths from a to b in Γ and from a' to b in Γ', preserving parity. Thus it is enough to prove the theorem for Γ'. Now for Γ' we have $e' = 2v'$, and this is the first reduction.

(ii) Let us define for a vertex w the two numbers

$$r(w) = \{\text{number of edges } s \text{ with } \varepsilon_1(s) = w\},$$

$$l(w) = \{\text{number of edges } s \text{ with } \varepsilon_2(s) = w\}.$$

Further, $f(w) = r(w) - l(w)$ will be called the flux and $r(w) + l(w)$ the order of w.

Assume now that Γ has at least one unicursal path $s_1, \ldots, s_h$ from a to b. We have $\varepsilon_2(s_i) = \varepsilon_1(s_{i+1})$, with $i = 1, \ldots, h - 1$ and we deduce:

(I) If $a = b$ then $f(w) = 0$ for all vertices.

(II) If $a \neq b$ then $f(w) = 0$ for all vertices except a and b, where $f(a) = 1$, and $f(b) = -1$. In this case, we add a new vertex and two edges as follows:

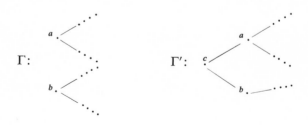

The graph Γ' has $e' = e + 2$ edges, $v' = v + 1$ vertices; we have again $2v' = e'$ and there is a 1–1 correspondence preserving parities between unicursal paths in Γ from a to b and unicursal paths in Γ' from c to c. Furthermore, in Γ' all vartices have flux zero. This is the second reduction.

Let us resume our reductions. We can consider only connected graphs Γ with $e = 2v$ and $f(w) = 0$ for all vertices. The initial and terminal point of the considered unicursal paths coincide; we call this point the base point.

We start the proof under these hypotheses. We make some constructions of new graphs and it is necessary to check that all such constructions yield new graphs verifying always the same hypotheses.

First assume that in the graph we have a configuration:

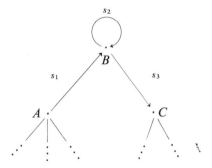

If B is the base point any unicursal path either starts or ends with s_2. Exchanging the role of s_2 as initial or terminal edge, we have a 1–1 correspondence between odd and even paths (since e is an even number). If the initial point is not B then every unicursal path contains $s_1 s_2 s_3$. We lump the three edges s_1, s_2, s_3 in one canceling the vertex B:

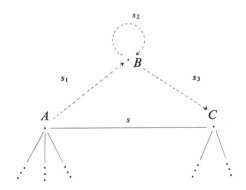

A unicursal path in the new graph Γ' is just a unicursal path in Γ with $s_1 s_2 s_3$ substituted by s.

The graph Γ' has $v' = v - 1$, $e' = e - 2$ vertices and edges, where $2v' = e'$ and the 1–1 correspondence between paths in Γ and Γ' preserves the parity. The graph Γ' has fewer vertices than Γ and so this case is done by induction.

Next, assume that Γ contains a configuration of the type

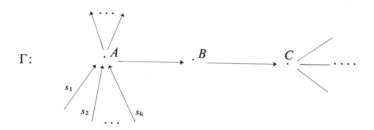

with the order of A greater than 2. It is clear, under our hypotheses, that if there is a vertex of order 2 there is also one with this further property.

Say we have k edges $s_1, \ldots, s_k$ with terminal point A. For each $i = 1, \ldots, k$ we construct a new graph Γ_i as follows:

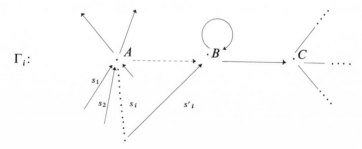

The graph Γ_i satisfies all the hypotheses and falls in the previous case. Each unicursal path in Γ yields a unicursal path in Γ_i for some i and conversely; the parity is preserved and so we are done by the previous case.

We now assume that the two previous cases do not apply; we use the hypothesis $2v = e$. We see that $\sum_w r(w) + l(w) = 2e = 4v$ and, therefore, for at least one w, we have $r(w) + l(w) \leqslant 4$. As we have excluded the case of vertices with order 2, the order of every vertex is 4. Since we have also excluded the configuration

the graph must contain the configuration

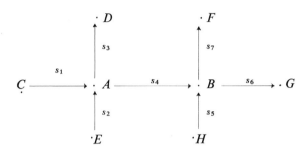

We proceed as in the previous case and we construct the graphs Γ_1, Γ_2;

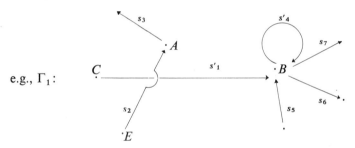

e.g., Γ_1:

Any unicursal path of Γ containing $s_1 s_4$ gives rise to a unicursal path in Γ_1, and one containing $s_2 s_4$ to a unicursal path in Γ_2; the parity is preserved. Now there are in Γ_1 (or Γ_2, respectively) some paths which do not come from paths of Γ, explicitly these containing $s_1' s_7$ and $s_5 s_4' s_7$ ($s_2' s_6$ and $s_5 s_4' s_7$, respectively); they are exactly those paths unicursal on the graphs Γ_6, Γ_7;

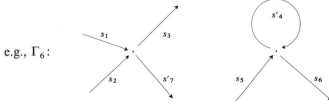

e.g., Γ_6:

The parities are again preserved.

As the theorem is true for Γ_1, Γ_2, Γ_6, and Γ_7 by the previous cases, it is true for Γ.

§7   The Standard Identity for Matrices

We go back now to **5.2.**

Proof of Theorem **5.2**   Since S_{2m} is multilinear, it is enough to show that it vanishes when computed on the elements of a basis of $(\Lambda)_m$. We choose the usual basis formed by the matrix units e_{ij}; for simplicity we write (i, j) instead of e_{ij}. Let us, therefore, choose $2m$ matrix units

$$\xi_1 = (i_1, j_1), \xi_2 = (i_2, j_2), \ldots, \xi_{2m} = (i_{2m}, j_{2m})$$

and compute $S_{2m}(\xi_1, \ldots, \xi_{2m})$; if two of the ξ_i coincide we already know that this value is zero. We assume, therefore, that the $2m$ chosen units are distinct. We construct a graph whose vertices are the numbers $1, 2, \ldots, m$ and whose edges are the $2m$ pairs previously written; we set in the obvious way $\varepsilon_1(i, j) = i$, $\varepsilon_2(i, j) = j$. A product $\xi_{\sigma(1)}\xi_{\sigma(2)} \cdots \xi_{\sigma(2m)}$ is nonzero if and only if it represents a unicursal path in this graph, and it yields the value (i, j) if and only if it starts with i and ends in j. Under the hypothesis of Theorem **6.3**, therefore, (i, j) is obtained an equal number of times from even and from odd permutations. Therefore, in the sum $\sum \varepsilon(\sigma)\xi_{\sigma(1)} \cdots \xi_{\sigma(2m)}$ the two contributions are canceled. It follows that $S_{2m}(\xi_1, \ldots, \xi_{2m}) = 0$ and the theorem is fully proved.

We can distinguish the proper subalgebras of a matrix algebra by the property that they satisfy a fixed polynomial identity.

Theorem 7.1   Let Λ be a field, m a natural number, and $x_j^{(i)}$ a variable with $i = 1, \ldots, m, j = 1, \ldots, 2m - 2$. The polynomial

$$S_{2m-2}(x_1^{(1)}, \ldots, x_{2m-2}^{(1)})S_{2m-2}(x_1^{(2)}, \ldots, x_{2m-2}^{(2)}) \ldots .$$
$$S_{2m-2}(x_1^{(m)}, \ldots, x_{2m-2}^{(m)})$$

does not vanish on $(\Lambda)_m$ but vanishes on any proper subalgebra $R \subset (\Lambda)_m$.

Proof   We have seen that

$$S_{2m-2}(e_{11}, e_{12}, e_{22}, e_{23}, \ldots, e_{m-1\ m}) = e_{1m}.$$

Similarly, we can substitute the variables to obtain e_{ij}, $i \neq j$. Therefore, we can substitute the variables in the product polynomial in order to get $e_{12}e_{21}e_{12}e_{21} \cdots \neq 0$.

Now let $R \subset (\Lambda)_m$ and $R \neq (\Lambda)_m$. Let Ω be an algebraic closure of Λ. We pass to $\bar{R} = R \otimes \Omega \subset (\Omega)_m$. Thus $\bar{R} \neq (\Omega)_m$. Let J be the radical of $\bar{R}$. We have $J^m = 0$ and $\bar{R}/J \simeq \oplus_i (\Omega)_{k_i}$. Since $\dim \bar{R} < m^2$ we must have $k_i < m$ for all i; therefore S_{2m-2} vanishes on $\bar{R}/J$. This means that S_{2m-2} evaluated in $\bar{R}$ gives elements in J. Since $J^m = 0$ we see that the polynomial under consideration vanishes on $\bar{R}$ and so also on R.

§8   Generalized Polynomial Identities

Most of the categorical results of the previous paragraph can be extended to a much more general situation.

We consider again $_{\Lambda}\mathscr{A}$, the category of Λ algebras. Let R be a fixed algebra. Consider the following category, which we call $_R/\mathscr{A}$: The objects of $_R/\mathscr{A}$ are Λ maps $R \to S$, with $S \in {_{\Lambda}\mathscr{A}}$, and morphisms are commutative diagrams:

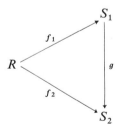

This category has many properties in common with the category of all algebras. It has infinite products since, if $R \xrightarrow{f_\alpha} S_\alpha$ is a family of objects, $R \xrightarrow{f_\alpha} \prod S_\alpha$ is a product. It has free objects, in fact, if $\mathscr{I}$ is a set and $\Lambda\{x_i\}_{i \in \mathscr{I}}$ the free algebra over Λ in the x_i's, one can form the free product $R * {_\Lambda}\Lambda\{x_i\}$ and one has a map $R \to R *_\Lambda \Lambda\{x_i\}$. By the universal properties of free products this is a free object in variables x_i in this category, in the sense that the maps from $R \to R *_\Lambda \Lambda\{x_i\}$ into $R \to S$ correspond to families $s_i \in S$, $i \in \mathcal{J}$. For some purposes it is necessary to generalize the concept of free algebras and consider the following situation. If M is a Λ module and $R \xrightarrow{f} S \in {}_R/\mathscr{A}$ we consider the set $\mathrm{Hom}_\Lambda(M, S)$, with S considered as a Λ module, and we seek an element of ${}_R/\mathscr{A}$ representing this functor. Then it is easy to see that if $T_\Lambda(M)$ indicates the tensor algebra of M, the object $R \xrightarrow{i} R * T_\Lambda(M)$ (where $R *_\Lambda T_\Lambda(M)$ is the free product in the category of Λ algebras and i is the canonical map) is endowed with a particular Λ map $J: M \to R * T(M)$ which is universal in the following sense. If $R \xrightarrow{f} S \in {}_R/\mathscr{A}$ and

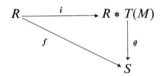

is a morphism, the correspondence $g \to gj$ is bijective between such morphisms and elements of $\mathrm{Hom}_\Lambda(M, S)$.

One can generalize in the obvious way the concept of variety in this category and the results of §§1 and 2 carry over to this case.

If $R = \Lambda$ then ${}_\Lambda/\mathscr{A}$ is just the category of Λ algebras and we obtain again the theory already developed as a special case.

We make the formalism a little more explicit. If R is a Λ algebra and M a Λ module, consider as before,

$$R * T(M) \simeq \bigoplus_{i=0}^\infty \{[\otimes_i (R \otimes M)] \otimes R\} \simeq \bigoplus_{i=0}^\infty \{R \otimes [\otimes_i (R \otimes M)]\}$$

The algebra $R * T(M)$ is equipped with two canonical maps, $i: R \to R * T(M)$ a Λ-algebra homomorphism and $j: M \to R * T(M)$ a Λ module homomorphism. If $M = \bigoplus_{i=0} \Lambda x_i$ is a free module over the basis x_i then $T(M) = \Lambda\{x_i\}$, the free algebra, and $R * T(M) = \bigoplus R \otimes x_{i_1} \otimes R \otimes x_{i_2} \otimes \cdots \otimes x_{i_s} \otimes R$. One usually drops the tensor signs and writes an element of $R * \Lambda\{x_i\}$ as $\sum r_1 x_{i_1} r_2 x_{i_2} \dots r_s x_{i_s} r_{s+1} = f(x_i)$.

If $f: R \to S$ is in ${}_R/\mathscr{A}$, we consider the maps in ${}_R/\mathscr{A}$ between $i: R \to R * T(M)$ and $f: R \to S$; we call them R maps and indicate their set by $\mathscr{M}_R(R * T(M), S)$. We know that

$$\mathscr{M}_R(R * T(M), S) \simeq \mathrm{Hom}_\Lambda(M, S)$$

canonically. Again, if $M = \oplus \Lambda x_i$ is free then $\mathrm{Hom}_\Lambda(M, S) \simeq S^I$ canonically.

The evaluation map is $\mu: R * T(M) \times \mathscr{M}_R(R * T(M), S) \to S$, where $\mu(u, \varphi) = \varphi(u)$, and so by the canonical isomorphism $\mathscr{M}_R(R * T(M), S) \simeq \mathrm{Hom}_\Lambda(M, S)$ is a map $\mu: R * T(M) \times \mathrm{Hom}_\Lambda(M, S) \to S$.

If $M = \oplus \Lambda x_i$ is free then $\mathrm{Hom}_\Lambda(M, S) \simeq S^I$; one usually writes $\mu(u(x_i), (s_i)) = u(s_i)$, the evaluation of the polynomial $u(x_i)$ in the "point" s_i.

By exponentiation the evaluation map gives rise to a map

$$\lambda: R * T(M) \to S^{\,\mathrm{Hom}_\Lambda(M,S)}$$

this is a ring homomorphism and, therefore, we have two interesting objects associated with λ, $\ker \lambda$ and $\mathrm{Im}\, \lambda$.

Definition 8.1   The ideal $\ker \lambda$ is called the ideal of generalized polynomial identities of S (in the variables $m \in M$) with coefficients in R. The ring $\mathrm{Im}\, \lambda$ is called the ring of generalized polynomial functions of S (in the variables $m \in M$) with coefficients in R.

If M, N are two Λ modules, $\mathscr{P}(M, N)$ denotes the module of polynomial maps. If N is an algebra, also $\mathscr{P}(M, N)$ is an algebra. It is easily checked that the map λ previously considered takes values in $\mathscr{P}(\mathrm{Hom}_\Lambda(M, S), S)$.

Usually, generalized polynomial identities do not give too stringent information on the rings under consideration unless further restrictions are imposed. In particular, if R is not a prime ring then there exist two non-zero elements $a, b \in R$ with $aRb = 0$, i.e., the ring R satisfies the nontrivial generalized polynomial identity axb with coefficients in R; so it seems that for nonprime rings the mere existence of a generalized polynomial identity is not a restriction. On the other hand, a theorem of Martindale characterizes in a rather tight form those prime rings which satisfy a nontrivial (in a sense to be specified) generalized polynomial identity [62].

§9   Generalized Versus Ordinary Polynomial Identities

As we remarked in the previous paragraph, we do not have in general a very stringent relationship between generalized and ordinary polynomial identities; we study here a case in which such relations are instead very strong. It is the case in which R is an Azumaya algebra.

Definition 9.1

(1) If $f: R \to S$ is a map of rings we set

$$S^R = \{s \in S \mid f(r)s = sf(r), \forall r \in R\}$$

as the centralizer of R in S.

(2) More generally, if Q is an R bimodule we set

$$Q^R = \{q \in Q \mid rq = qr, \forall r \in R\}.$$

We recall the basic property of Azumaya algebras. If $f: R \to S$ is a map of Λ algebras and R is an Azumaya algebra over Λ, then the canonical homomorphism $\bar{f}: R \otimes_\Lambda S^R \to S$ is an isomorphism, where $\bar{f}$ is given by $\bar{f}(r \otimes s) = f(r)s$.

Then let R be an Azumaya algebra over Λ, S a U algebra, $f: R \to S$ an algebra map, and M a Λ module. We consider the setup of the previous paragraph and form $R \to R *_\Lambda T(M)$, and the evaluation

$$\lambda: R *_\Lambda T(M) \to \mathscr{P}(\text{Hom}_\Lambda(M, S), S).$$

We shall now study this situation.

Theorem 9.2

(1) $(R * T(M))^R$ is isomorphic to $T(N)$ where $N = (R \otimes M \otimes R)^R$.

(2) $R * T(M) \simeq R \otimes T(N)$.

Proof   Consider $R \otimes T(N)$; we construct a map $\mu: R \otimes T(N) \to R * T(M)$ in the following way. We have the map $R \xrightarrow{i} R * T(M)$ sending R in the "constants," and we have a module map $N \to R \otimes M \otimes R \to R * T(M)$. Therefore, by universality we obtain a map $\varphi: T(N) \to R * T(M)$. Since $N \subseteq (R * T(M))^R$ we have that $\varphi(T(N)) \subseteq (R * T(M))^R$ and so we can define a ring homomorphism $R \otimes T(N) \to R * T(M)$. On the other hand, we define a map $v: R * T(M) \to R \otimes T(N)$ by the universal property of free products. We map $R \to R \otimes T(N)$ in the canonical way and we map $M \to R \otimes T(N)$ by composing the inclusion map $M \to R \otimes M \otimes R$

with the inverse isomorphism of $R \otimes N \to R \otimes M \otimes R$ (given by the fact that R is an Azumaya algebra, $R \otimes M \otimes R$ a bimodule, and $N = (R \otimes M \otimes R)^R$). Therefore we have in this way a map $T(M) \to R \otimes T(N)$ and finally $v\colon R * T(M) \to R \otimes T(N)$. We claim that μ and v are inverse isomorphisms, which will finish the proof. To verify that μ and v are inverse isomorphisms we notice that $R * T(M)$ is generated as an algebra by $R \oplus R \otimes M \otimes R$ and $R \otimes T(N)$ is generated by $R \oplus R \otimes N$. It is easy to see that, by the very construction of μ and v, they give inverse isomorphisms of the two modules $R \oplus R \otimes M \otimes R$ and $R \oplus R \otimes N$ and, therefore, μ and v are inverse isomorphisms of the two algebras.

Next we have a similar theorem for the other object of our study.

Theorem 9.3

(1) $\mathscr{P}(\mathrm{Hom}_\Lambda(M, S), S)^R$ is isomorphic to $\mathscr{P}(\mathrm{Hom}_\Lambda(N, S^R), S^R)$.

(2) $\mathscr{P}(\mathrm{Hom}_\Lambda(M, S), S) \simeq R \otimes \mathscr{P}(\mathrm{Hom}_\Lambda(N, S^R), S^R)$.

Proof If E is any Λ module, consider $\mathscr{P}(E, S)$: we have a canonical map of rings $S \to \mathscr{P}(E, S)$ sending the elements of S in the constant polynomial functions, and so also a map $R \to S \to \mathscr{P}(E, S)$. Since R is an Azumaya algebra, we have $\mathscr{P}(E, S) \simeq R \otimes \mathscr{P}(E, S)^R$. It is clear that a polynomial map $\gamma \in \mathscr{P}(E, S)^R$ if and only if the image of γ is in S^R; therefore, $\mathscr{P}(E, S)^R = \mathscr{P}(E, S^R)$. We must study $\mathscr{P}(\mathrm{Hom}_\Lambda(M, S), S^R)$. We claim that $\mathrm{Hom}_\Lambda(M, S)$ is canonically isomorphic to $\mathrm{Hom}_\Lambda(N, S^R)$; in fact,

$$\mathrm{Hom}_\Lambda(M, S) \simeq \mathrm{Hom}_\Lambda(M, R \otimes S^R)$$

$$\mathrm{Hom}_\Lambda(M, \mathrm{Hom}_\Lambda(R^*, S^R)) \simeq \mathrm{Hom}_\Lambda(M \otimes R^*, S^R).$$

But now we have an algebra isomorphism $\tau\colon R \otimes_\Lambda R^o = R^e \to \mathrm{End}_\Lambda(R)$ given by the formula $\tau(r \otimes s)(u) = rus$ (R^o is the opposite of R and R^e the enveloping algebra), and under this isomorphism, $(R \otimes_\Lambda R)^R$ corresponds to $\mathrm{Hom}_\Lambda(R, \Lambda)$ with $\Lambda \subseteq R$ the center of R. Therefore,

$$\mathrm{Hom}_\Lambda(M \otimes R^*, S^R) \simeq \mathrm{Hom}_\Lambda(M \otimes (R \otimes_\Lambda R)^R, S^R).$$

Finally, we need only to remark that $M \otimes (R \otimes_\Lambda R)^R \simeq (R \otimes M \otimes R)^R$; this follows from the general remark that if R is a finitely generated projective R^e module, an R bimodule Q is nothing else but an R^e module and $Q^R \simeq \text{Hom}_{R^e}(R, Q) \simeq \text{Hom}_{R^e}(R, R^e) \otimes_{R^e} Q$. So in general, if M is a Λ module, then

$$(Q \otimes_\Lambda M)^R \simeq \text{Hom}_{R^e}(R, R^e) \otimes_{R^e} (Q \otimes M)$$

$$\simeq (\text{Hom}_{R^e}(R, R^e) \otimes_{R^e} Q) \otimes_\Lambda M \simeq Q^R \otimes_\Lambda M.$$

We can finally collect all our information in a diagram. We have the isomorphism $\nu: R * T(M) \to R \otimes T(N)$, the isomorphism just constructed, that we will call π. Then $\pi: \mathscr{P}(\text{Hom}_\Lambda(M,S), S)^R \to \mathscr{P}(\text{Hom}_\Lambda(N, S^R), S^R)$. We have, further, the two evaluation maps

$$\lambda: R * T(M) \to \mathscr{P}(\text{Hom}_\Lambda(M, S), S)$$

$$\text{and} \quad \tilde{\lambda}: T(N) \to \mathscr{P}(\text{Hom}_\Lambda(N, S^R), S^R);$$

Therefore, we can construct the following diagram:

$$
\begin{array}{ccc}
R * T(M) & \xrightarrow{\;\lambda\;} & \mathscr{P}(\text{Hom}_\Lambda(M, S), S) \simeq R \otimes \mathscr{P}(\text{Hom}_\Lambda(M, S), S)^R \\
\simeq \downarrow {\scriptstyle \nu} & & \simeq \downarrow {\scriptstyle 1_R \otimes \Pi} \\
R \otimes T(N) & \xrightarrow{\;1_R \otimes \tilde{\lambda}\;} & R \otimes \mathscr{P}(\text{Hom}_\Lambda(N, S^R), S^R).
\end{array}
$$

Theorem 9.4   The above diagram is commutative.

Proof   It is enough to check the commutativity on the generators $R \oplus R \otimes M \otimes R$ of $R * T(M)$ and this last one follows from carefully checking all the maps under consideration.

Basically the theorem tells us that if we want to study generalized polynomial identities or generalized polynomial maps with coefficients in an Azumaya algebra, we are reduced to study ordinary polynomial identities and polynomial maps. In fact, under the identification $R * T(M) \simeq R \otimes T(N)$ we have $\ker \lambda \equiv R \otimes \ker \tilde{\lambda}$, and under the identification

$$\mathscr{P}(\mathrm{Hom}_\Lambda(M, S), S) \equiv R \otimes \mathscr{P}(\mathrm{Hom}_\Lambda(N, S^R), S^R)$$

we have $\mathrm{Im}\ \lambda \equiv R \otimes \mathrm{Im}\ \tilde{\lambda}$. In particular, we can analize the following case. Assume that $S^R = \Gamma$ is a commutative Λ algebra (in particular, if $S = R$ we have $S^R = \Lambda$), then $\mathrm{Hom}_\Lambda(N,\ \Gamma) \simeq \mathrm{Hom}_\Gamma(N \otimes_\Lambda \Gamma,\ \Gamma)$. We have a map from $S(N)$, the symmetric algebra over N, into $\mathscr{P}(\mathrm{Hom}_\Lambda(N, \Gamma), \Gamma)$ given by sending

$$N \to \mathrm{Hom}_\Lambda(\mathrm{Hom}_\Lambda(N, \Gamma), \Gamma) \subseteq \mathscr{P}(\mathrm{Hom}_\Lambda(N, \Gamma), \Gamma)$$

canonically. The map $\tilde{\lambda} \colon T(N) \to \mathscr{P}(\mathrm{Hom}_\Lambda(N,\ \Gamma),\ \Gamma)$ factors canonically through

$$T(N) \to S(N) \to \mathscr{P}(\mathrm{Hom}_\Lambda(N, \Gamma), \Gamma).$$

Therefore, to study $\tilde{\lambda}$ we have to study separately the two maps $T(N) \to S(N)$ and $S(N) \to \mathscr{P}(\mathrm{Hom}_\Lambda(N,\ \Gamma),\ \Gamma)$. Under a suitable hypothesis this second map induces an isomorphism ϕ; for instance, assume Λ to be an infinite field and M a finite-dimensional vector space, then N is also a finite-dimensional vector space and $S(N) \otimes_\Lambda \Gamma \to \mathscr{P}(\mathrm{Hom}_\Lambda(N, \Gamma), \Gamma)$ is an isomorphism. In particular,

$$\mathrm{Im}\ \tilde{\lambda} \otimes \Gamma = \mathscr{P}(\mathrm{Hom}_\Lambda(N, \Gamma), \Gamma)$$

and $\ker \tilde{\lambda}$ is the commutator ideal of $T(N)$, and so in this case we have a complete description of the objects under consideration. Even more concretely, let $R = (\Lambda)_m$, the ring of $m \times m$ matrices over Λ, and $M = \sum_{i=1}^n \Lambda x_i$ be free over the x_i's. Choose the usual canonical basis e_{ij} for $(\Lambda)_m$. We see in this case that $N = ((\Lambda)_m \otimes M \otimes (\Lambda)_m)^{(\Lambda)^m}$ is the subspace spanned by the linearly independent elements $\xi_{ij,t} = \sum_{s=1}^m e_{si} x_t e_{js}$. The various objects that we have considered are in this case described more concretely as follows: $T(M) = \Lambda\{x_i\}$, $T(N) = \Lambda\{\xi_{ij,t}\}$ the free algebra in the respective variables, $(\Lambda)_m \otimes T(N) = (\Lambda\{\xi_{ij,t}\})_m$ the ring of $m \times m$ matrices over $\Lambda\{\xi_{ij,t}\}$. Under the identification

$$(\Lambda)_m * \Lambda\{x_i\} \simeq (\Lambda)_m * T(M) \simeq (\Lambda)_m \otimes T(N) \simeq (\Lambda\{\xi_{ij,t}\})_m$$

we see that the variable x_t is identified with the matrix $\sum_{ij} \xi_{ij,t} e_{ij}$, which is the generic matrix with entries being the free variables $\xi_{ij,t}$. Our theorem reads that if $f \in (\Lambda)_m * T(M)$ then f is a generalized polynomial identity of $(\Lambda)_m$ if and only if, considering f as a matrix in $(\Lambda\{\xi_{ij,t}\})_m$, all the entries of such a matrix vanish when considered as commutative polynomials; finally, the generalized polynomial function in n variables $((\Lambda)_m)^n \to (\Lambda)_m$ are just all the ordinary polynomial maps between the two vector spaces.

Chapter II

STRUCTURE THEOREMS

§1   Primitive PI-Algebras

In this paragraph we prove the first main structure theorem for PI-algebras.

Theorem 1.1 (Kaplansky)   Let R be a primitive Λ algebra satisfying a nontrivial identity $f(x)$ of degree d. Then R is a central simple algebra of dimension n^2 over its center Z, with $2n \leqslant d$.

Proof   We may assume that $f(x)$ is multilinear. Let M be a faithful irreducible module, D the centralizer of this module, and K a maximal subfield of D. We identify R with the ring of endomorphisms of M which R induces and we form the ring

$$RK = \{\sum r_i k_i | r_i \in R, k_i \in K\} \subseteq \mathrm{Hom}_K(M, M).$$

The module M is an irreducible RK module and its centralizer is K. The ring RK still satisfies $f(x)$ since $f(x)$ is multilinear. We also notice that Λ is naturally mapped into K so that we may think of the coefficients of $f(x)$ as being in K. We claim that $2\dim_K M \leqslant d$, which will prove that $RK \simeq \mathrm{Hom}_K(M, M)$ and

$$\dim_Z R = \dim_K \mathrm{Hom}_K(M, M) = n^2,$$

n $= \dim_K M$, and $2n \leqslant d$.

Assume then that we had a finite dimensional subspace W of M with $2\dim_K W > d$. We have, by Jacobson's density theorem, a subalgebra T of RK which induces on W the full ring of endomorphisms $\mathrm{Hom}_K(W, W)$. The algebra T satisfies $f(x)$ so that $\mathrm{Hom}_K(W, W)$ would satisfy $f(x)$; this is a contradicition of **5.1** of Chapter I.

Corollary 1.2   Let R be a Λ algebra satisfying an identity $f(x)$ of degree d for which $c(f) = \Lambda$. If M is an irreducible module, D the centralizer, and Z the center of D, we have

$$\dim_D M = k < \infty, \qquad \dim_Z D = h^2 < \infty, \qquad 2hk \leqslant d.$$

Proof   The proof follows from the previous theorem where we notice that the assumption $c(f) = \Lambda$ insures that for every ideal I of R the identity $f(x)$ is nontrivial for R/I.

Corollary 1.3   Let R be a Λ algebra satisfying an identity $f(x)$ for which $c(f) = \Lambda$. Then every primitive ideal of R is maximal.

§2   Prime Rings

We want to develop here some elementary results on prime rings for which we have no uniform reference and which will be used in the subsequent discussion of our structure theorems.

Proposition 2.1   Let R be a prime ring, and $I \subseteq R$ a left ideal. If $r(I) = \{r \in R | Ir = 0\}$, then $r(I) \cap I$ is a two-sided ideal of I and $I/r(I) \cap I$ is a prime ring.

Proof   It is obvious that $r(I) \cap I$ is a two-sided ideal of I. Now let $a, b \in I$ be such that $aIb \subseteq r(I) \cap I$. This means that $IaIb = 0$.

Ia and Ib are two left ideals of R whose product is 0. Since R is a prime ring we must have either $Ia = 0$ or $Ib = 0$, i.e., either $a \in r(I) \cap I$ or $b \in r(I) \cap I$.

Proposition 2.2   Let R be a prime Λ algebra.
(1) $M = \text{Ann}(R)$ is a prime ideal of Λ.
(2) R considered as a Λ/M module is torsion free.
(3)   If K is the field of fractions of Λ/M, then $R \otimes_\Lambda K$ is still a prime ring containing R.

Proof   (1)   Let α, $\beta \in \Lambda$ be such that $\alpha\beta \in M$, i.e., $\alpha\beta R = 0$. Then $\alpha R\beta R = 0$. Since αR and βR are ideals of R, we must have either $\alpha R = 0$ or $\beta R = 0$, i.e., $\alpha \in M$ or $\beta \in M$.
(2)   Let $\alpha \in \Lambda$. Consider $I_\alpha = \{r \in R | \alpha r = 0\}$. I_α is clearly a two-sided ideal of R. Furthermore, $\alpha R I_\alpha = 0$; therefore, either $\alpha R = 0$ or $I_\alpha = 0$.
(3)   The ring $R' = R \otimes_\Lambda K$ contains R since R is torsion free over Λ/M. Let a', $b' \in R'$ with $a'R'b' = 0$, where $a' = a\alpha^{-1}$, $b' = b\beta^{-1}$, a, $b \in R$, and α, $\beta \in \Lambda/M$. Therefore, $aRb = 0$ so that either $a = 0$ or $b = 0$ which implies either $a' = 0$ or $b' = 0$.

Proposition 2.3   Let R be a prime Λ algebra satisfying a nontrivial identity. Then R has no nonzero nil left ideals.

Proof   In virtue of **2.2** we can pass to $S = R \otimes_\Lambda K$ which satisfies an identity with coefficients in K, a field. If N is a nil left ideal of R, then NK is a nil left ideal of S. Therefore, we now work in S. Let us assume that S has a nonzero nil left ideal I. We can clearly assume that $I = Sa$, $a^2 = 0$. Let

$$f(x_1, \ldots, x_n) = x_1 g(x_2, \ldots, x_n) + h(x_1, \ldots, x_n)$$

be a multilinear identity for S, where h is the sum of all monomials not starting with x_1 and the variables are chosen in such a way that $g(x_2, \ldots, x_n)$ is a nonzero polynomial. Set $x_1 = ar, x_2 = r_2 a, \ldots, x_n = r_n a$. Since $a^2 = 0$, we have $h(ar, r_2 a, \ldots, r_n a) = 0$. Therefore $0 = f(ar, r_2 a, \ldots, r_n a) = arg(r_2 a, \ldots, r_n a)$. Since S is prime $a \neq 0$ and r is arbitrary, we must have $g(r_2 a, \ldots, r_n a) = 0$ for all choices of $r_2, \ldots, r_n \in S$. Therefore $g(x_2, \ldots, x_n)$ is a multilinear identity of degree $n - 1$ in the prime algebra

$I/r(I) \cap I$ (see **2.1**). It is nontrivial since the coefficients are in the field K. We can now use an induction on the degree of the identity and deduce that $I/r(I) \cap I$ has no nonzero nil ideals. Now $I/r(I) \cap I$ is nil; therefore, we must have $I/r(I) \cap I = 0$, i.e., $I \subseteq r(I)$. This implies $I^2 = 0$, a contradiction since S is prime.

We deduce some information on nil ideals in general, provided that the rings under consideration satisfy proper identities.

Lemma 2.4   Let R be a Λ algebra satisfying a proper identity $f(x)$. If $L(R)$ denotes the lower nil radical of R we have $L(R) = \{\cap \ P | P$ prime ideal such that $f(x)$ is nontrivial on $R/P\}$.

Proof   Let $r \notin L(R)$; we have to show that we can find a prime ideal P such that $r \notin P$ and $f(x)$ is nontrivial on R/P. Now $r \notin L(R)$ if and only if it is not strongly nilpotent, i.e., there exist two sequences ˙

$$r = r_1, r_2, \ldots, r_n, \ldots \quad \text{and} \quad c_1, c_2, \ldots, c_n, \ldots \in R$$

such that $r_n = r_{n-1} c_{n-1} r_{n-1}$ and $r_n \neq 0$ for all n.

Let $\alpha_1, \ldots, \alpha_t$ be the coefficients of $f(x)$ and $c(f) = (\alpha_1, \ldots, \alpha_t)$ the ideal generated by the α_i. If $s \in R$ and $s \neq 0$ we have $c(f)s \neq 0$ and therefore $c(f)^m s \neq 0$ for all m. Now if we had a k such that $\alpha_i^k s = 0$ for all i, we would have $c(f)^{tk+1} s = 0$, a contradiction. Therefore, there is an i such that $\alpha_i^m s \neq 0$ for all m. We apply this to all the elements r_n and since $\alpha_i^m r_n = 0$ implies $\alpha_i^m r_k = 0$ for $k > n$ we deduce the existence of an i such that $\alpha_i^m r_n \neq 0$ for all m and n. Consider then the sequence $r_n' = \alpha_i^{2^{n-1}} r_n$. We have $r_1' = \alpha_i r$, $r_n' = r_{n-1}' c_{n-1} r_{n-1}'$, and $r_n' \neq 0$; therefore, $\alpha_i r$ is not strongly nilpotent. This implies that we can find a prime ideal P with $\alpha_i r \notin P$. In R/P we have $\alpha_i \bar{r} \neq 0$, therefore $r \notin P$ and $f(x)$ is nontrivial in R/P.

Corollary 2.5   If R satisfies a proper identity $f(x)$, then the upper nil radical and the lower nil radical of R coincide; moreover any right or left nil ideal is contained in $L(R)$.

Proof   By Proposition **2.3**, if N is a nil left ideal, then $N \subseteq P$ for any prime ideal P such that the identity $f(x)$ is nontrivial in R/P. By **2.4** these ideals intersect in $L(R)$. The same reasoning applies symmetrically for nil right ideals.

§3   Structure of Semiprime Rings

Theorem 3.1 (Amitsur)   Let R be a Λ algebra satisfying a proper identity of degree d. Let $L(R)$ be the lower nil radical of R, then

(1)   There exists a finite number of commutative Λ algebras $\Lambda_1, \ldots, \Lambda_s$, where $s \leqslant d/2$, each without nilpotent ideals, and an embedding

$$0 \rightarrow R/L(R) \rightarrow \underset{i}{\oplus} (\Lambda_i)_i$$

such that all the stable identities (in particular, the multilinear ones) satisfied by R are also satisfied by $\oplus_i(\Lambda_i)_i$.
(2)   If R is unitary, this embedding preserves 1.
(3)   If R is a prime ring, there is a field K and an embedding

$$0 \rightarrow R \rightarrow (K)_s$$

with $s \leqslant d/2$ and $(K)_s = KR$

Proof   (1)   Consider first the case in which R is prime. We embed R in $S = R \otimes_\Lambda K$ (symbols as in **2.2**) and then S in $S[x]$, the polynomial ring. The ring $S[x]$ clearly satisfies all the stable identities satisfied by R. Furthermore, since S has no nil ideals (**2.3**), $S[x]$ is semisimple ([44], p. 150): $J(S[x]) = 0 = \{\cap \mathcal{M} | \mathcal{M}$ primitive ideals$\}$. The ring $S[x]$ satisfies a multilinear identity with coefficients in K (a field) of degree $\leqslant d$; therefore, we can apply the main structure theorem (**1.1**) and we see that if $\mathcal{M}$ is a primitive ideal of $S[x]$, then $S[x]/\mathcal{M} \simeq (D)_k$, and if Z is the center of D we have $\dim_Z D = h^2$ and $2hk \leqslant d$. Letting $\Omega_\mathcal{M}$ be an algebraic closure of Z we see that $D \otimes_Z \Omega_\mathcal{M} \simeq (\Omega_\mathcal{M})_{n_\mathcal{M}}$, $n_\mathcal{M} = hk$. We have a map

$$S[x] \rightarrow S[x]/\mathcal{M} \rightarrow (\Omega_\mathcal{M})_{n_\mathcal{M}}$$

with kernel $\mathcal{M}$. Furthermore $(\Omega_\mathcal{M})_{n_\mathcal{M}}$ still satisfies all stable identities satisfied by R. Letting $\mathscr{M}$ vary we get an injection

$$j: R \to S[x] \to \prod_{\mathscr{M}} S[x]/_{\mathscr{M}} \to \prod (\Omega_{\mathscr{M}})_{n_{\mathscr{M}}}.$$

Let $\Lambda_i = \prod_{n_{\mathscr{M}} = i} \Omega_{\mathscr{M}}$, $i \leqslant d/2$; Λ_i has no nilpotent elements. We have $\prod_{n_{\mathscr{M}} = i} (\Omega_{\mathscr{M}})_{n_{\mathscr{M}}} = (\Lambda_i)_i$ and $\prod (\Omega_{\mathscr{M}})_{n_{\mathscr{M}}} \simeq \oplus (\Lambda_i)_i$ and j is the desired embedding.

In the general case $L(R) = \bigcap P$, P running on the prime ideals such that $f(x)$ is nontrivial on R/P. We embed $R/P \to \prod (\Omega_{\mathscr{M}})_{n_{\mathscr{M}}}$, and then

$$0 \to R/L(R) \to \prod R/P \to \prod\prod (\Omega_{\mathscr{M}})_{n_{\mathscr{M}}};$$

again grouping together the factors relative to the matrices of the same size, we have the desired result.

(2)   This statement follows immediately since, when R has a 1, all morphisms under consideration preserve 1.

(3)   Let $j: R \to \oplus_{i \leqslant d/2} (\Lambda_i)_i$ be the embedding given before and assume that R is prime. Set

$$\varphi_i: R \to \oplus (\Lambda_i)_i \xrightarrow{\pi_i} (\Lambda_i)_i$$

where π_i is the ith projection of the direct sum. Let $N_i = \ker \varphi_i$. We have $\bigcap_{i \leqslant d/2} N_i = 0$ since j is an embedding. Now R is a prime ring so we must have $N_{i_0} = 0$ for one index and therefore for this index $\varphi_{i_0}: R \to (\Lambda_{i_0})_{i_0}$ is a monomorphism. For simplicity we write this monomorphism

$$\varphi: R \to (\Lambda)_i, \ i \leqslant d/2.$$

Now let I be an ideal of Λ maximal with respect to the following property: $R \to (\Lambda)_i \to (\Lambda/I)_i$ is a monomorphism. Such an I exists by Zorn's lemma. We claim that I is a prime ideal. Then let $U \supseteq I$ and $V \supseteq I$ be two-sided ideals such that $UV \subseteq I$. Consider

$$\psi: R \to (\Lambda)_i \to (\Lambda/U)_i \text{ and } \gamma: R \to (\Lambda)_i \to (\Lambda/V)_i.$$

We have $\ker \psi \ker \gamma = 0$; therefore, as R is prime, we must have that one of the two is 0, say $\ker \psi = 0$. In this case, using the maximality of U, we deduce that $U = I$. This proves that I is prime.

Letting K be the quotient field of Λ/I, consider the map

$$\lambda: R \to (\Lambda)_i \to (\Lambda/I)_i \to (K)_i.$$

λ is an embedding and $(K)_i$ satisfies all the multilinear identities satisfied by R. To finish the argument we have to show that $\lambda(R)K = (K)_i$. Now $\lambda(R)K$ is a subalgebra of $(K)_i$; if it were a proper subalgebra of $(K)_i$ it would satisfy a multilinear identity not satisfied by $(K)_i$ (Chapter I, **7.1**). Therefore we would have a contradiction, unless $\lambda(R)K = (K)_i$.

The last part admits a converse.

Theorem 3.2   A ring R is a prime ring with a nontrivial polynomial identity if and only if there exists a monomorphism $\lambda: R \to (K)_i$ with K a field and $(K)_i = \lambda(R)K$.

Proof   In one direction it is the content of part (3) of the previous theorem. Let now $\lambda: R \to (K)_i$ be as in the statement of the theorem; we identify R with its image $\lambda(R)$ for simplicity of notation. We must show that R is a prime ring with nontrivial polynomial identities. The ring R satisfies all the identities satisfied by $(K)_i$ so also the identity S_{2i} which is nontrivial on R. It remains to be proved that R is prime. Letting $a, b \in R$ be such that $aRb = 0$, we have $a(K)_i b = aRKb = aRbK = 0$. Since $(K)_i$ is prime we must have $a = 0$ or $b = 0$ and R is also prime.

§4   Consequences of the Structure Theorem

The main consequence is a theorem which allows us to forget about proper identities, nontrivial identities, and so on. In fact, until now we always had to be careful that a given identity be nontrivial on some algebra under consideration; this would clearly be avoided simply if our identities had coefficients ± 1.

Theorem 4.1 (Amitsur)   If R is a Λ algebra satisfying a proper identity $f(x)$, then R satisfies $S_n(x_1, \ldots, x_n)^m$ for suitable n, m. In particular, it satisfies a multilinear identity with coefficients ± 1.

Proof   Let $\Lambda\{x_s\}$ be a free algebra in sufficiently many variables and I the T ideal of polynomial identities of R. The ring $S = \Lambda\{x_s\}/I$ satisfies $f(x)$ and this identity is proper for S (Chapter I, **3.6**). From **3.1** we have an embedding

$$S/L(S) \to (\Lambda_i)_i.$$

Now the ring $(\Lambda_i)_i$ satisfies the standard identity $S_d(x_1, \ldots, x_d)$. Consider d variables among the x_s's. We can still call them $x_1, \ldots, x_d$, with $\tilde{x}_i$ their classes in S and $\bar{x}_i$ their classes in $S/L(S)$.

We have $S_d(\bar{x}_1, \ldots, \bar{x}_d) = 0$ and therefore $S_d(\tilde{x}_1, \ldots, \tilde{x}_d) \in L(S)$. The ideal $L(S)$ is nil so that for some m we have $S_d(\tilde{x}_1, \ldots, \tilde{x}_d)^m = 0$. Therefore $S_d(x_1, \ldots, x_d)^m \in I$, where I is the ideal of all identities of R. Therefore S_d^m is a polynomial identity of R. If we multilinearize this polynomial we get a polynomial with coefficients ± 1.

Corollary 4.2   If R satisfies a proper identity then it is a *PI* algebra (in the sense of **3.5**)

From now on we will tacitly use the fact that a *PI* algebra satisfies multilinear identities with coefficients ± 1.

Another corollary of secondary importance is the following.

Proposition 4.3   Let R be a *PI* ring with 1. Assume that $a, b \in R$ are elements such that $ab = 1$. Then $ba = 1$.

Proof   Consider $u = 1 - ba$. If P is a prime ideal of R we have a monomorphism $i\colon R/P \to (K)_i$, where K is a field. Since $ab = 1$ we have $\bar{a}\bar{b} = 1$ in $\bar{R} = R/P$. Since $(K)_i$ is a matrix algebra over a field it follows that $\bar{b}\bar{a} = 1$ in $\bar{R}$, so $u \in P$. Since P was arbitrary, $u \in \bigcap P = L(R)$, so u is nilpotent, $u^k = 0$ for some k, and

$$u^k = (-ba + 1)^k = \sum_{i=0}^{k} \binom{k}{i}(-ba)^i = 1 + (\sum_{i=1}^{k} \binom{k}{i}(-ba)^{i-1}b)a.$$

Therefore $1 = ca$ and

$$c = -\sum_{i=1}^{k} \binom{k}{i}(-ba)^{i-1}b.$$

Now $b = cab = c$ so $ba = 1$.

§5 The Structure of Prime Rings

We will give a very precise structure theorem for prime PI rings as a consequence of **3.1** and a careful study of regular elements.

Lemma 5.1   Let R be a PI algebra, $R \neq 0$. Let $c \in R$ be an element. There is a two sided ideal $I \neq 0$ and a number k such that $I \subseteq Rc + l(c^k)$ where $l(r) = \{a \in R | ar = 0\}$ for $r \in R$.

Proof   Each algebra Rc^k satisfies a multilinear identity; we choose one of minimal degree d_k. Since $Rc^{k+1} \subseteq Rc^k$ we have $d_{k+1} \leqslant d_k$. Therefore, we can find a k such that $d_{k+1} = d_k$; call this integer d for simplicity and let $g(x_1, \ldots, x_d)$ be the identity of degree d satisfied by Rc^k. We can assume $d \geqslant 2$, otherwise $Rc^k = 0$, so $R = l(c^k)$ and the lemma is proved.

Choosing the name of the variables suitably we can assume that

$$g(x_1, \ldots, x_d) = g_1(x_2, \ldots, x_d)x_1 + \sum_{i=2}^{n} h_i(x_1, \ldots, \check{x}_i, \ldots, x_d)x_i$$

with $g_1(x_2, \ldots, x_d)$ a nonzero polynomial.

Let $T = \{g_1(r_2 c^{k+1}, \ldots, r_d c^{k+1}) | r_i \in R\}$. We have $T \neq 0$ as $d_{k+1} = d_k = d$ and so g_1 is not an identity of Rc^{k+1}. Now $TR \subseteq Rc + l(c^k)$ since

$$0 = g(rc^k, r_2 c^{k+1}, \ldots) = g_1(r_2 c^{k+1}, \ldots, r_d c^{k+1})rc^k + \sum_{i=2}^{n} h_i(\cdots)r_i c^{k+1}$$

and so

$$(g_1(r_2 c^{k+1}, \ldots, r_d c^{k+1})r + uc)c^k = 0$$

for some u. Therefere

$$g_1(r_2 c^{k+1}, \ldots, r_d c^{k+1})r + uc \in l(c^k)$$

with r arbitrary.

Now clearly $T \subseteq Rc$ so the two-sided ideal I generated by T is nonzero and contained in $Rc + l(c^k)$.

Corollary 5.2   If $c \in R$ and $l(c) = 0$ there is a two-sided ideal $I \neq 0$ with $I \subseteq Rc$.

We use this lemma to characterize regular elements in a prime PI ring.

Theorem 5.3   Let R be a prime PI ring.

(1)   An element $a \in R$ is not a left zero divisor if and only if Ra contains a nonzero two-sided ideal (similarly on the right).

(2)   If $R \subseteq (K)_i$, with K a field, and $RK = (K)_i$ then $a \in R$ is not a left zero divisor in R if and only if it is invertible in $(K)_i$.

(3)   If a is not a left zero divisor then it is not a right divisor and thus it is regular.

(4)   If $a = bc$ is a regular element, b and c are regular.

Proof   (1)–(3). We have seen that if $l(a) = 0$ then $Ra \supseteq I$, $I \neq 0$, which is a two-sided ideal. Assume conversely that $Ra \supseteq I$, I is a nonzero two-dided ideal, and embed R in $(K)_i$ with the property of (2). We claim that $KI = (K)_i$. In fact $(K)_i$ is a simple algebra so that

$$(K)_i = (K)_i KI(K)_i = KRKIKR \subseteq KI$$

since I is a two-sided ideal of R. Therefore, $(K)_i a = KRa \supseteq KI = (K)_i$ so that a is invertible in $(K)_i$. This proves (1)–(3) at the same time.

(4)   This follows again from (2) and the basic properties of the matrix algebra $(K)_i$.

We need one further lemma.

Lemma 5.4   Let $R \subseteq (K)_i$ be a ring with K a field and $RK = (K)_i$. Then there is an $a \in R$ invertible in $(K)_i$.

Proof   In characteristic 0 the proof is almost trivial. Let

$$a_1, \ldots, a_{i^2} \in R$$

be a basis of $(K)_i$ over K. The polynomial $\det(\sum x_j a_j)$ is not identically zero so there are integers $n_1, \ldots, n_{i^2}$ with $\det(\sum n_j a_j) \neq 0$, therefore, $\sum n_j a_j \in R$ is invertible in $(K)_i$.

In general we proceed as follows. For $a \in R$, let $l_K(a) = \{u \in (K)_i | ua = 0\}$. Then $l(a) = l_K(a) \cap R$. Now $l_K(a)$ is a left ideal of $(K)_i$ and thus a subspace of the finite dimensional vector space $(K)_i$. Therefore, the set of left ideals of R, $\{l(a)\}$, satisfies the ascending and descending chain condition. This gives us a sublemma.

Sublemma   Let J be a nonzero left ideal of R; there is an $a \in J$ with $a \neq 0$ and $l(a) = l(a^2)$. For such an a we have $Ra \cap l(a) = 0$.

Proof   Clearly for every $b \in J$, we have $l(b) \subseteq l(b^2) \subseteq l(b^3) \subseteq \ldots$ and by the chain condition we have $l(b^n) = l(b^{2n})$ for some n. If we always had $b^n = 0$ for every $b \in J$ and suitable n, we would have that J is a nil ideal; this is absurd since R is a prime PI ring. Therefore, we can find an element $a = b^n$, with $a \neq 0$ and $l(a) = l(a^2)$. If $r \in Ra \cap l(a)$ we have $r = sa$ and $saa = 0$; therefore, $s \in l(a^2) = l(a)$, so $sa = 0$.

We now prove Lemma **5.4**.

Let $a \in R$ be such that $l(a) = l(a^2)$ and $l(a)$ is minimal in the set of ideals with this property (we can apply the descending chain condition). We must show $l(a) = 0$ which will imply that a is invertible in $(K)_i$ by **5.3**. Assume that $l(a) \neq 0$ and $j \in l(a)$ is such that $j \neq 0$ and $l(j) = l(j^2)$ (sublemma). Consider $a + j$. We claim that $l(a + j) = l((a + j)^2)$. First, $l(a) \cap l(j) = l(a + j)$, since if $s(a + j) = 0$ we have $sa = -sj \in Ra \cap l(a) = 0$ therefore $sa = sj = 0$ and so $s \in l(a) \cap l(j)$. Conversely, it is clear that $l(a) \cap l(j) \subseteq l(a + j)$ so that $l(a) \cap l(j) = l(a + j)$.
Similarly,

$$l((a + j)^2) = l(a^2 + (a + j)j) = l(a^2) \cap l((a + j)j) = l(a) \cap l(aj + j^2);$$

but if $s \in l(a) \cap l(aj + j^2)$, then $0 = s(aj + j^2) = sj^2$; therefore,

$$l(a) \cap l(aj + j^2) = l(a) \cap l(j^2) = l(a) \cap l(j).$$

Finally, $l(a + j) = l(a) \cap l(j) = l((a + j)^2)$.

Since a was chosen so that $l(a)$ would be minimal with respect to $l(a) = l(a^2)$, we must have $l(a + j) = l(a)$, i.e., $l(a) \subseteq l(j)$; hence $j \in l(a) \subseteq l(j)$ making $j^2 = 0$ a contradiction.

Corollary 5.5   Let R be a prime PI ring, with $I \neq 0$ a two-sided ideal. Then I contains a regular element of R.

Proof   Let $R \subseteq (K)_i$, with K a field, and $RK = (K)_i$. We know that $IK = (K)_i$; therefore, the hypotheses of **5.4** apply and there is an $a \in I$, with a invertible in $(K)_i$. Clearly a is regular in R.

Corollary 5.6   Let P be a prime ideal in a PI ring R. Let $S = \{a \in R | a$ is regular modulo $P\}$. Then P is maximal with respect to the exclusion of S.

Proof   Let $I \supset P$ be a two-sided ideal. Consider $\bar{I} = I/P \subseteq R/P$. $\bar{I}$ is a nonzero two-sided ideal and, therefore, contains a regular element of R/P, hence $\bar{I} \cap S \neq \phi$.

We pass now to the main theorem.

Theorem 5.7 (Posner)   Let R be a prime PI algebra.

(1) R has a total ring of left and right fractions $Q(R)$.
(2) $Q(R)$ is a simple-algebra finite dimensional over its center $Z(R)$.
(3) $Q(R) = RZ(R)$.
(4) $Q(R)$ satisfies all the polynomial identities satisfied by R.
(5)   If $R \subseteq (K)_i$ with $(K)_i = KR$ we can complete the diagram

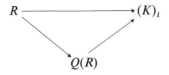

and $(K)_i \simeq K \otimes_{Z(R)} Q(R)$.

Proof   (1) Let R be prime, $R \subseteq (K)_i$, K a field, and $(K)_i = KR$. Let S be the set of regular elements of R; if $a \in S$, then a is invertible in $(K)_i$, so we can consider the sets

$$Q_l(R) = \{a^{-1}b \mid a, b \in R, \quad a \in S\}$$
$$Q_r(R) = \{ba^{-1} \mid a, b \in R, \quad a \in S\}.$$

If we prove that $Q_l(R)$ and $Q_r(R)$ are rings then they are clearly the total left and right rings of fractions of R and they coincide. By left–right symmetry it is enough that we look at $Q_l(R)$. To show that it is a ring it is enough to verify the Öre condition, i.e., if a, $b \in R$ and $a \in S$ then there are c, $d \in R$, with $cb = da$ and $c \in S$. Consider then Ra. We can find **(5.3)** a nonzero two-sided ideal $I \subseteq Ra$, By **5.5** we can find a $c \in I$, $c \in S$. It follows that $cb \in I \subseteq Ra$; therefore, $cb = da$ for a suitable $d \in R$.

(2) and (5) We claim that $Q(R)$ is simple; clearly $Q(R)K = (K)_i$. Let $I \neq 0$ be a two-sided ideal of $Q(R)$. By **5.5** we can find an $a \in I$ invertible in $(K)_i$. Now $a \in Q(R)$ so $a = bc^{-1}$, b, $c \in R$, $c \in S$. If a is invertible in $(K)_i$ we have $b \in S$ so $a^{-1} = cb^{-1} \in Q(R)$ and $I = Q(R)$. Let $Z(R)$ be the center of $Q(R)$. Since $Q(R)K = (K)_i$ we must have $Z(R) \subseteq K$ (center of $(K)_i$). Now the canonical morphism $Q(R) \otimes_{Z(R)} K \to (K)_i$ is surjective and as $Q(R) \otimes_{Z(R)} K$ is simple ([44], p. 90) the map is also injective. Thus $Q(R) \otimes_{Z(R)} K \to (K)_i$ is an isomorphism. In particular, $\dim_{Z(R)} Q(R) = i^2$; (2) and (5) are proved.

(3) Consider $Z(R)R \subseteq Q(R)$. Since $(K)_i = KR \cong K \otimes_{Z(R)} Z(R)R$, we must have $Z(R)R = Q(R)$.

(4) The algebra $Q(R)$ is a finite-dimensional vector space over $Z(R)$ that we consider as an algebraic variety over $Z(R)$. Let $\bar{R}$ be the Zariski closure of R in $Q(R)$. We claim that $\bar{R} = Q(R)$. In fact, let $a \in R$ be an invertible element in $Q(R)$. We have $aR \subseteq R$. Therefore, since multiplication by a is continuous in the Zariski topology we have $a\bar{R} \subseteq \bar{R}$; hence $\bar{R} \supseteq a\bar{R} \supseteq a^2\bar{R} \supseteq a^3\bar{R} \ldots$, a descending chain of closed sets (multiplication by a

is a homeomorphism). By the descending-chain condition for closed sets, in a variety we have $a^n \bar{R} = a^{n+1} \bar{R}$ for some n.

Since a is invertible we have $a^{-1} \bar{R} = \bar{R}$. It follows that $\bar{R} = Q(R)$ as every element of $Q(R)$ is of the form $a^{-1}b$, with $a, b \in R$.

Since $\bar{R} = Q(R)$, every polynomial function vanishing on R vanishes on $Q(R)$; in particular, every polynomial identity of R is a polynomial identity of $Q(R)$.

It will be useful in the future to consider an intermediate ring between R and $Q(R)$ (with R a prime PI ring). It is defined as follows: $Q(R)$ is a simple algebra so we have the reduced trace map $\mathrm{Tr}: Q(R) \to Z(R)$. We consider the ring U generated by $\mathrm{Tr}(R)$. The main fact that we will need is the following.

Proposition 5.8 $Z(R)$ is the field of fractions of U.

Proof   Let $z \in Z(R)$, we have that $z = a^{-1}b$, with $a, b \in R$. Hence $z\mathrm{Tr}(a) = \mathrm{Tr}(b)$. If we knew that $\mathrm{Tr}(a) \neq 0$ we would be done. In general we know that $\mathrm{Tr}(Q(R)) \neq 0$, and as $Q(R) = RZ(R)$ and $Q(R) = aQ(R)$, we deduce

$$\mathrm{Tr}(Q(R)) = \mathrm{Tr}(aQ(R)) = \mathrm{Tr}(aRZ(R)) = Z(R)\mathrm{Tr}(aR) \neq 0.$$

Therefore there is a $c \in R$ such that $\mathrm{Tr}(ac) \neq 0$. Then from $az = b$ we deduce $(ac)z = (bc)$ and $z = \mathrm{Tr}(bc)/\mathrm{Tr}(ac)$.

Corollary 5.9   Let R be a prime PI algebra over a commutative ring A, with $Q(R)$ its ring of fractions. If R is algebraic over A then $Q(R)$ is also algebraic over A.

Proof   Let $Z(R)$ be the center of $Q(R)$. It is enough to show that $Z(R)$ is algebraic over A. Considering U as before we must prove only that U is algebraic over A. Now if $r \in R$, and $\sum_{i=0}^{m} \alpha_i r^i = 0$ with $\alpha_i \in A$, we have $\sum_{i=0}^{m} \alpha_i \lambda^i = 0$ for all eigenvalues of r, therefore $\mathrm{Tr}(r)$, which in a sum of eigenvalues, is also algebraic over A.

§6   Extension Maps and Spectrum

We make a digression here on some auxiliary ideas of a very general nature; no polynomial identities will appear until the end of the digression.

Definition 6.1   If R is a ring we define the following.

(1) $\text{Spec}(R) = \{P | P \text{ a prime ideal of } R\}$.
(2) If $S \subset R$ is a subset we set $V(S) = \{P \in \text{Spec}(R) | S \subset P\}$.

Warning   If we work in the category of rings (without 1) it is advantageous to consider $R \subseteq R$ as a prime ideal. In the category of rings with 1 instead, a prime ideal P will always be $\neq R$.

Proposition 6.2
(1) $V(\bigcup S_\alpha) = \bigcap V(S_\alpha)$.
(2) $V(SRT) = V(S) \cup V(T)$.
(3) $V(\phi) = \text{Spec}(R), V(R) = \phi$.

Proof   (1)   This part is clear.
(2)   If $P \in V(S) \cup V(T)$, then $P \supset S$ or $P \supset T$ so that $P \supset SRT$ and $P \in V(SRT)$. Conversely, assume that $P \supset SRT$. If $P \not\supset S$, there is an element $a \in S$, $a \notin P$. Since $P \supset SRT$, we have for every $b \in T$ the relation $aRb \subset P$, and as P is prime, $b \in P$. Therefore, $P \supset T$. This proves that $V(SRT) \subset V(S) \cup V(T)$ hence $V(SRT) = V(S) \cup V(T)$.
(3)   This is clear.

The properties proved in **6.2** tell us that the sets $V(S)$ can be considered as the closed sets of a topology, called the Zariski topology of $\text{Spec}(R)$.

Remark   If $\varphi \colon R \to S$ is a map of rings, we do not necessarily have an induced map $\varphi^* \colon \text{Spec}(S) \to \text{Spec}(R)$ (as we would in the commutative case), because a subring of a prime ring is not necessarily prime. This suggests the following construction.

Given a map $\varphi: R \to S$, we define:

$$S^R = \{x \in S \mid x\varphi(r) = \varphi(r)x, \text{ for all } r \in R\}.$$

We consider S as an R bimodule via φ and we write rs, sr instead of $\varphi(r)s$, $s\varphi(r)$.

Further, recall that if A, B are two subrings of C and $B \subseteq C^A$, then $A \subseteq C^B$ and the set

$$AB = \{\sum_i a_i b_i \mid a_i \in A, b_i \in B\}$$

is a subring of C.

Definition 6.3

(1)   A map $\varphi: R \to S$ is called an extension if $S = \varphi(R)S^R$.

(2)   A map $\varphi: R \to S$ is called a finitely generated extension if there exists a finite set of elements $x_1, \ldots, x_k \in S^R$ such that $S = \varphi(R)\{x_1, \ldots, x_k\}$, where $\{x_1, \ldots, x_k\}$ denotes the subring generated by the elements $x_1, \ldots, x_k$.

(3)   In the previous definitions, if we replace S^R with $Z(S) = $ center of S, we will speak respectively of central extensions and finitely generated central extensions.

Proposition 6.4   Rings with 1 and extension maps are a subcategory of the category of rings with 1.

Proof   The identity map $1_R: R \to R$ is an extension since $1 \in R^R$. Next one has to show that if $\varphi: R \to S$ and $\psi: S \to T$ are extension maps, then $\psi\varphi$ is an extension map. The proof is trivial.

The main application of the concepts defined is now given.

Theorem 6.5   Let $\varphi: R \to S$ be an extension.

(1)   If $P \in \text{Spec}(S)$ then $\varphi^{-1}(P) \in \text{Spec}(R)$.

(2)   The map $\varphi^*: \text{Spec}(S) \to \text{Spec}(R)$ given by $\varphi^*(P) = \varphi^{-1}(P)$ is continuous.

(3)   We obtain in this way a controvariant functor from the category of rings with 1 and extensions to topological spaces.

Proof (1) Let $P \in \text{Spec}(S)$ and a, $b \in R$ be such that $aRb \subset \varphi^{-1}(P)$. Then

$$\varphi(a)S\varphi(b) = \varphi(a)\varphi(R)S^R\varphi(b) = \varphi(a)\varphi(R)\varphi(b)S^R = \varphi(aRb)S^R \subseteq PS^R \subseteq P,$$

hence either $\varphi(a) \in P$ or $\varphi(b) \in P$. Therefore, either $a \in \varphi^{-1}(P)$ or $b \in \varphi^{-1}(P)$, and so $\varphi^{-1}(P)$ is a prime ideal.

(2) It is clear that $\varphi^{*-1}(V(U)) = V(\varphi(U))$ for every $U \subseteq R$, so φ^* is continuous.

(3) It is obvious that $(\varphi\psi)^* = \psi^*\varphi^*$.

Proposition 6.6 Let $\varphi: R \to S$ be an extension. If I is a two-sided ideal of R, then $\varphi(I)S$ and $S\varphi(I)$ are two-sided ideals of S. If R has a 1 then $\varphi(I)S = S\varphi(I)$.

Proof If R has a 1 then

$$\varphi(I)\varphi(R)S^R = \varphi(IR)S^R = \varphi(I)S^R = S^R\varphi(I) = S^R\varphi(RI)$$
$$= S^R\varphi(R)\varphi(I) = S\varphi(I)$$

and, since $S\varphi(I) = \varphi(I)S$, it is clear that it is a two-sided ideal. Otherwise $\varphi(I)S$ is clearly a right ideal and

$$S\varphi(I)S = S^R\varphi(R)\varphi(I)S = S^R\varphi(RI)S \subseteq \varphi(I)S^RS \subseteq \varphi(I)S.$$

Proposition 6.7 Let $i: R \to S$ be an inclusion map and an extension.
(1) If U, V denote the centers of R and S, we have $U \subseteq V$.
(2) If S is prime, R is prime.
(3) If S is semiprime, R is semiprime.

Proof (1) Given $S = RS^R$; if $r \in U$ we have that r commutes with all the elements of R. Also $r \in R$, so r commutes with all the elements of S^R and it follows that r commutes with all the elements of S.

(2) S prime means that (0) is a prime ideal, so by **6.5**, we have $(0) = i^{-1}(0)$ is a prime ideal of R.

(3) S semiprime means $(0) = \{\bigcap P | P$ prime ideal of $S\}$; hence $(0) = \{\bigcap i^{-1}(P) | P$ prime in S, and $i^{-1}(P)$ is a prime ideal of $R\}$.

As stated at the beginning, we apply the preceding theory to *PI* algebras.

Theorem 6.8   Let $i: R \to S$ be an inclusion extension of *PI* algebras and S a prime ring; then we have the following.

(1) R is a prime ring.

(2) $a \in R$ is regular in R if and only if it is regular in S.

(3)   There is a canonical commutative diagram of inclusion extension maps

where $Q(R)$, $Q(S)$ are the rings of fractions of R and S.

(4)   If $Z(R)$, $Z(S)$ are the centers of $Q(R)$ and $Q(S)$, respectively, and $T = Q(S)^{Q(R)}$, we have $Z(R) \subseteq Z(S)$ and a canonical isomorphism $Q(R) \otimes_{Z(R)} T \simeq Q(S)$.

(5)   If S^R is the center of S then $T = Z(S)$.

Proof   (1)   has been proved in **6.7.**

(2)   If $a \in R$ is regular in S then it is clear that it is regular in R. If a is regular in R we use **5.3** and deduce the existence of a nonzero two-sided ideal $I \subseteq Ra$. Then $Sa \supset SI$ which is a nonzero two-sided ideal by **6.6**; therefore, **5.3** applies again and a is regular in S.

(3)   Consider the map $R \to S \to Q(S)$. Every regular element of R is regular in S, hence invertible in $Q(S)$; therefore, by the universal property of rings of fractions, we must have a map $j: Q(R) \to Q(S)$ making the diagram

commutative. Since $Q(R)$ is a simple ring, j must be an inclusion. Now all maps are extensions except possibly j; but $R \to S \to Q(S)$ is an extension, being a composition of extensions, and $Q(S) = (RS^R)Z(S)$. Now every element of $Q(R)$ is of the form ab^{-1}, with $a, b \in R$, so if $u \in Q(S)$ centralizes R it will also centralize ab^{-1} and hence $Q(R)$. Therefore $S^R Z(S) \subseteq Q(S)^{Q(R)}$ and $Q(R) \to Q(S)$ is an extension.

(4)   Since $j: Q(R) \to Q(S)$ is an extension we have $Z(R) \subseteq Z(S)$ (by **6.7**). We have a homomorphism

$$\lambda: Q(R) \otimes_{Z(R)} T \to Q(S)$$

which is onto since j is an extension. On the other hand, the kernel of λ, being an ideal of $Q(R) \otimes_{Q(R)} T$, must be of the form $Q(R) \otimes_{Z(R)} I$ with I an ideal of T. But $T \subseteq Q(S)$; therefore, $I = 0$.

(5)   We have $Q(R) = RS^R Z(R)$; if S^R is the center of S then $S^R \subseteq Z(S)$, so that $Q(S) = RZ(S)$ and, therefore, $Q(S) = Q(R)Z(S)$ and $T = Z(S)$.

§7   Irreducible Representations

In this paragraph we want to make more explicit the results of **5.6** and to recast them in a new form in which they acquire a better light.

Proposition 7.1   Let R be a ring and K a field. For a map $\alpha: R \to (K)_i$ the following two conditions are equivalent.

(1)   For every field $H \supseteq K$ the set of matrices induced by the composite map $\alpha: R \to (K)_i \to (H)_i$ is irreducible.

(2) $\alpha(R)K = (K)_i$.

Proof   (1) $\Rightarrow$ (2)   Let $H \supseteq K$ be algebraically closed; then $\alpha(R) \subseteq (H)_i$ is irreducible if and only if $\alpha(R)H = (H)_i$, which implies $\alpha(R)K = (K)_i$.

(2) $\Rightarrow$ (1)   If $\alpha(R)K = (K)_i$, for every $H \supseteq K$ we have $\alpha(R)H = (K)_i H = (H)_i$ so $\alpha(R)$ is irreducible in $(H)_i$.

Definition 7.2   A map $\alpha: R \to (K)_i$ satisfying the condition of **7.1** will be called an absolutely irreducible representation of dimension i (AI representation for short).

Remark   The map $\alpha: R \to (K)_i$ is absolutely irreducible if and only if it is a central extension.

We want to establish an equivalence relation among such representations.

Definition 7.3   Given two AI representations $\alpha: R \to (K)_i$ and $\beta: R \to (L)_j$, we say that they are equivalent if $i = j$ and there exists a field M containing K and L and an M-automorphism ψ of $(M)_i$ such that the diagram

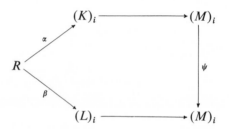

is commutative.

Remark   It is quite easy to see that we have in fact defined an equivalence relation in this way.

Theorem 7.4   Let R be an algebra.

(1)   If $\alpha: R \to (K)_i$ is an AI representation, ker α is a prime ideal and $R/\ker \alpha$ is a PI algebra.

(2)   Two AI representations α, β are equivalent if and only if ker α = ker β.

(3)   In this way we establish a 1–1 correspondence between equivalence classes of representations and prime ideals P of R such that R/P is a PI algebra.

Proof (1) The map $R/\ker \alpha \to (K)_i$ is a central inclusion so (1) follows from **3.2**.

(2) If α and β are equivalent it is clear that $\ker \alpha = \ker \beta$. Conversely, assume $\ker \alpha = \ker \beta = P$. Let $S = R/P$; we have two central inclusions, $\bar{\alpha}: S \to (K)_i$ and $\bar{\beta}: S \to (L)_j$. If $Q(S)$ denotes the quotient ring of S we have (by **5.7** or **6.8**) isomorphisms

$$\tilde{\alpha}: Q(S) \otimes_{Z(S)} K \to (K)_i, \qquad \tilde{\beta}: Q(S) \otimes_{Z(S)} L \to (L)_j$$

making the diagrams

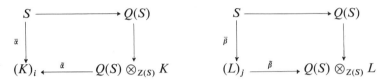

commutative. Therefore, $i^2 = \dim_{Z(S)} Q(S) = j^2$ so $i = j$. Further, let M be a field containing K and L and consider

$$\gamma: Q(S) \otimes_{Z(S)} M \simeq (Q(S) \otimes_{Z(S)} K) \otimes_K M \simeq (K)_i \otimes_K M \simeq (M)_i$$

$$\delta: Q(S) \otimes_{Z(S)} M \simeq (Q(S) \otimes_{Z(S)} L) \otimes_L M \simeq (L)_i \otimes_L M \simeq (M)_i.$$

If we set $\psi = \delta\gamma^{-1}$ we have that the diagram

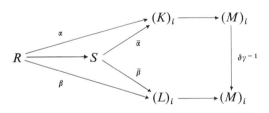

is commutative.

(3) We only have to prove that, if P is a prime ideal such that R/P satisfies a polynomial identity, there is an AI representation $\alpha: R \to (K)_i$ with $\ker \alpha = P$. This is just a restatement of **3.2**.

We end with some further definitions and remarks on the Zariski topology of R.

Definition 7.5

(1) $\text{Spec}_n(R) = \{P \in \text{Spec}(R) | P$ corresponds to an AI representation of dimension $n\}$.
$\sum_n (R) = \bigcup_{i \leqslant n} \text{Spec}_i(R)$.

(2) A ring R is said to be of degree $\leqslant m$ if $\text{Spec}(R) = \sum_m (R)$ and R is a PI algebra.

Remark   If R is a simple algebra then R is of degree m if and only if it is of dimension m^2 over its center.

Proposition 7.6

(1) $\sum_n (R)$ is a closed subset of $\text{Spec}(R)$.

(2) If R satisfies a proper polynomial identity of degree d then R is of degree $\leqslant [d/2]$ (with $[d/2]$ being the maximal integer $\leqslant d/2$).

Proof   (1) Let $\mathscr{V}$ be the variety generated by all rings of $n \times n$ matrices over fields. Let I be the minimal ideal of R such that $R/I \in \mathscr{V}$, (Chapter I, **1.4**). We claim that $\sum_n (R) = V(I)$. In fact, if $P \in \sum_n (R)$ then R/P can be embedded in $i \times i$ matrices for some $i \leqslant n$; therefore, it can be embedded in $n \times n$ matrices over some field and $R/P \in \mathscr{V}$ so that $P \subseteq I$. Conversely, if $P \subseteq I$ then R/P satisfies all the identities of $n \times n$ matrices; in particular $S_{2n}(x_1, \ldots, x_{2n})$. We have, therefore, an AI representation $\varphi: R \to (K)_i$ with kernel P, and $(K)_i$ satisfies S_{2n}; therefore, $i \leqslant n$ and so by definition $P \in \sum_n (R)$.

(2) If R satisfies a proper identity $f(x)$ of degree d, by Theorem **3.1** one has that $R/L(R)$ satisfies all the identities of $n \times n$ matrices for $n = [d/2]$; if $P \in \text{Spec}(R)$, then R/P satisfies the same identities. If $R \to (K)_i$ is an AI representation with kernel P, we must have that $(K)_i$ satisfies all multilinear identities of $n \times n$ matrices, so $i \leqslant n = [d/2]$ and $P \in \sum_{[d/2]}(R)$.

§8   Nil Subalgebras

We already have a good hold on nil ideals in a PI algebra by virtue of Chapter I, **2.5**. One can actually considerably strengthen those results, which is the purpose of this paragraph.

We recall the definition of $L_\alpha(R)$ with α an ordinal number. The ideal $L_\alpha(R)$ is defined inductively by the following properties:

(1) $L_1(R) = \{\sum M \mid M$ nilpotent ideal of $R\}$.
(2) $L_\alpha(R) = \bigcup_{\beta < \alpha} L_\beta(R)$ if α is a limit ordinal.
(3) $L_{\alpha+1}(R) \supseteq L_\alpha(R)$ and $L_{\alpha+1}(R)/L_\alpha(R) = L_1(R/L_\alpha(R))$.

The main property is that $L(R) = \bigcup L_\alpha(R)$.

Theorem 8.1   If R satisfies a proper multilinear identity of degree d, we have

(1) Every nil left or right ideal of R is contained in $L(R)$.
(2) If B is a nil subalgebra of R, then $B^{[d/2]} \subseteq L_1(R)$.
(3) $L(R) = L_2(R)$.

Proof   (1) is the content of Chapter I, **2.5**.
(2) Let us assume first of all that B is nilpotent, $B^m = 0$. Let us set $U_{2i-1} = B^{n-i}RB^{i-1}$ and $U_{2i} = B^{n-i}RB^i$, with n an integer, $i = 1, \ldots, n$. If $h \leqslant 2n$ we have

$$U_1 U_2 U_3 \ldots U_h = (B^{n-1}R)^h B^{[h/2]};$$

furthermore, $U_j U_k \subseteq RB^n R$ if $j > k$.
Let us now choose n minimum with respect to the property that $RB^n R$ is nilpotent (there is one such since $RB^m R = 0$). Let us assume by contradiction that $n > d/2$, i.e., $n - 1 \geqslant [d/2]$, and let us use the polynomial identity. We put in evidence one monomial of it:

$$\alpha x_1 x_2 \ldots x_d + \gamma(x),$$

where $\gamma(x)$ is a sum of other monomials, either in the same variables $x_1, \ldots, x_d$ in different order or with other variables. We substitute for x_i some $u_i \in U_i$ and for the other variables 0. Then $\gamma(u_i) \in RB^n R$ and therefore $\alpha u_1 u_2 \ldots u_d \in RB^n R$. We have thus proved that

$$\alpha U_1 U_2 \ldots U_d = \alpha(B^{n-1}R)^d B^{[d/2]} \subseteq RB^n R.$$

Since $RB^n R$ is nilpotent it follows that $(\alpha(B^{n-1}R)^d B^{[d/2]})^s = 0$ for some s.

Since $[d/2] \leqslant n - 1$ this implies that $(\alpha(B^{n-1}R)^d B^{n-1})^s = 0$ and therefore $(\alpha RB^{n-1}R)^{(d+1)s} = 0$.

We repeat the argument for all the other coefficients of the identity and we finally find an exponent t such that $\alpha^t(RB^{n-1}R)^t = 0$ for all coefficients α of $f(x)$. Then as in **2.4** it follows, since $f(x)$ is proper, that $(RB^{n-1}R)^t = 0$. This contradicts the minimality of n, so $n \leqslant [d/2]$, and $B^{[d/2]}$ generates a nilpotent ideal.

In general, even if B is not nilpotent, every finitely generated subalgebra of B will be nilpotent since B is a *PI* algebra, and therefore, $L(B) = B$ (Chapter I, **2.5**). On the other hand, B is the union of its finitely generated subalgebras C; each one of these C is nilpotent and therefore $C^{[d/2]} \subseteq L_1(R)$. It follows that $B^{[d/2]} \subseteq L_1(R)$.

(3)   We have from (2) that $L(R)^{[d/2]} \subseteq L_1(R)$; therefore, a fortiori, $L(R) = L_2(R)$.

Remark   It is not possible, in general, to better the inclusion $L(R)^{[d/2]} \subseteq L_1(R)$.

Example   Let Λ be a commutative ring whose nil radical N is not nilpotent. Consider the algebra $(\Lambda)_n$ of $n \times n$ matrices. It satisfies the standard identity S_{2n} of degree $2n$. Let U be the subset of $(\Lambda)_n$ formed by the matrices $(\alpha_{ij}) \in (\Lambda)_n$ with $\alpha_{ij} \in N$ if $i \leqslant j$. One verifies immediately that U is a nil subalgebra of $(\Lambda)_n$. Consider U^{n-1}; it contains the element $e_{n\,n-1}e_{n-1\,n-2} \cdots e_{21} = e_{n1}$. Now if $z \in N$, then $ze_{1n} \in U$ and $ze_{1n}e_{n1} = ze_{11}$. The two-sided ideal generated by e_{n1} contains ze_{11} for every $z \in N$. The ring $\{ze_{11} | z \in N\}$ is isomorphic to N, which is not nilpotent; therefore, $U^{n-1} \not\subseteq L_1((\Lambda)_n)$.

Chapter III

THE IDENTITIES OF MATRIX ALGEBRAS

§1   Rings of Generic Matrices

We have seen among the consequences of Chapter I, **4.6** that for a commutative ring Λ, if we consider the polynomial ring $\Lambda[x_{t,ij}]$ with $t \in T$ and $i, j = 1, \ldots, n$, the $n \times n$ matrices $\xi_t \in (\Lambda[x_{t,ij}])_d$ and $\xi_t = (x_{t,ij})$, and the map

$$\lambda: \Lambda\{x_t\} \rightarrow (\Lambda[x_{t,ij}])_d$$

defined by $\lambda(x_t) = \xi_t$; then ker λ is the ideal of polynomial identities of $(R)_d$ where R is any generic commutative algebra.

We now consider the case in which Λ is an integral domain and K its quotient field, and we will denote by $K(x_{t,ij})$ the quotient field of $\Lambda[x_{t,ij}]$.

We need some lemmas to further describe the structure of the ring $\Lambda\{\xi_t\}$.

Lemma 1.1   Let K be a field and n an integer. There exists a division ring D with a center that is an infinite field $F \supseteq K$ such that

(1) $\dim_F D = n^2$,
(2) D can be generated as F algebra by two elements.

Proof   We use the Hilbert construction. First we fix a field $G \supseteq K$ with a K automorphism σ of order exactly n [e.g., $G = K(x_1, \ldots, x_n)$ and $\sigma(x_i) = x_{i+1}$ with indices taken mod n]. Let

$$D = \{ \sum_{i=m}^{\infty} \alpha_i x^i | \alpha_i \in G, m \in Z \},$$

the ring of Laurent series, where the multiplication is according to the rule $x^i \alpha = \sigma^i(\alpha) x^i$ with $\alpha \in G$. It is easy to verify that D is a division ring with center $F = \{ \sum_{i=m}^{\infty} \alpha_i (x^n)^i | \alpha_i \in G^{\sigma} \}$.

If $\lambda_0, \ldots, \lambda_{n-1}$ is a basis of G over G^{σ}, then $\lambda_i x^j$ with $i, j = 0, \ldots, n - 1$ is a basis of D over F. Therefore, $\dim_F D = n^2$. On the other hand, if λ is a primitive element of G over G^{σ}, one has the basis $\lambda^i x^j$ with $i, j = 0, \ldots, n - 1$, and D is generated by λ and x.

Lemma 1.2   Let S be a simple algebra, finite dimensional over its center Z. Then S can be generated by two elements over Z.

Proof   We distinguish two cases.

(i)   Assume Z is finite. In this case $S = (Z)_n$ for some n. We prove something more, i.e., that S can be generated by two elements over $Z/(p)$, the prime field of Z. Let $G \supseteq Z$ be a field with $[G:Z] = n$. Let $\alpha \in G$ be a generator of G over $Z/(p)$ and let σ be an automorphism of G for which $G^{\sigma} = Z$ (with σ of order n). We have $S = (Z)_n \simeq \text{Hom}_Z(G, G)$. Consider $\sigma \in \text{Hom}_Z(G, G)$ and $\alpha \in G \subseteq \text{Hom}_Z(G, G)$. By the linear independence of automorphisms we see that the monomials $\sigma^i \alpha^j$ with $i, j = 1, \ldots, n$ are independent over Z; therefore, they generate S over Z. Now α generates $G \supseteq Z$ over $Z/(p)$, so α, σ generate S over $Z/(p)$.

(ii)   Assume Z is infinite. Let $D \supseteq F \supseteq Z$ be as in Lemma **1.1** with F the center of D, and let $x^i \lambda^j$, with $i, j = 0, \ldots, n - 1$, be a basis of D over F. We want to show that if D can be generated by two elements over its center, it follows that S can also be so generated. Let $u_1, \ldots, u_{n^2}$ be a basis of S over Z. Let $a = \sum \alpha_i u_i$, $b = \sum \beta_j u_j \in S$ with $\alpha_i, \beta_j \in Z$. We want to show that we can determine the elements α_i, β_j's in such a way that $a^i b^j$ with $i, j = 0, \ldots, n - 1$ are a basis of S over Z. Now given an n^2-tuple of elements of S, $c_1, \ldots, c_{n^2}$, the condition that they form a basis is $\det(\text{Tr}(c_i c_j)) \neq 0$ where Tr denotes the reduced trace in the algebra S. If we write this condition for the elements $a^i b^j$ we obtain a polynomial $f(\alpha_i, \beta_j) = \det(\text{Tr}(a^i b^j)(a^k b^s))$ and as Z is infinite it will be enough to show that this polynomial is formally nonzero. Consider an algebraically closed field $\Omega \supseteq F$; we have $D \otimes_F \Omega \simeq (\Omega)_n \simeq S \otimes_Z \Omega$. $u_1, \ldots, u_{n^2}$ gives a basis of $(\Omega)^d$ and the polynomial $f(\alpha_i, \beta_j)$ is formally nonzero if and only if we can find two elements $a, b \in (\Omega)_n$ such that $a^i b^j$ with $i, j = 0, \ldots, n - 1$ is a basis of $(\Omega)_n$ over Ω; now this can be clearly done, in fact $x, \lambda \in D \subseteq (\Omega)_n$ will do. The lemma is completely proved.

We go back to the ring of generic matrices $\Lambda\{\xi_t\}$; we assume that the set T of indices has at least two elements. With just one generic matrix ξ, we have that $\Lambda\{\xi\}$ is the polynomial ring in one variable.

Theorem 1.3   Let Λ be an integral domain.

(1) $\Lambda\{\xi_t\}$ has no zero divisors.
(2) $\Lambda\{\xi_t\}$ is an order in a division ring $\Lambda\langle\xi_t\rangle$ of dimension n^2 over its center.
(3) $\Lambda\{\xi_t\}$ is generic in the variety generated by $n \times n$ matrices.

Proof   (1) We first need a slight variation of Chapter I, **4.5.** Let K be the field of fractions of Λ. If S_1 and S_2 are as in Chapter I, **4.5** relative to the K under consideration, then they are equivalent over K. Since $\Lambda \subseteq K$ they are also equivalent over Λ (Chapter I, **2.13**). Now to construct $\Lambda\{\xi_t\}$ we choose a generic commutative Λ algebra Γ and we perform the construction of Chapter I, **4.6** relative to Γ and $(\Gamma)_n$. We can now choose Γ to be an infinite field containing K. If D is a division ring of dimension n^2 over its center $Z \supseteq K$ then, by what we have remarked, D and $(\Gamma)_n$ are equivalent; therefore, to find the identities of $n \times n$ matrices, we may apply the construction of Chapter I, **4.6** to D and consider $E = Z[x_{t,i}] \otimes_Z D$, $t \in T$ with $i = 1, \ldots, n^2$, $u_1, \ldots, u_{n^2}$ a basis of D and $\xi_t' = \sum x_{t,i} u_i \in E$. The ring $\Lambda\{\xi_t'\}$ is again isomorphic to the free algebra modulo the identities of $n \times n$ matrices, so $\Lambda\{\xi_t'\} \simeq \Lambda\{\xi_t\}$. But now clearly E is a domain and so (1) is proved.

(2) $\varphi : \Lambda\{\xi_s\} \to (\Lambda[x_{t,ij}])_n \to (K(x_{t,ij}))_n$ is an inclusion. If ξ_1, ξ_2 are two distinct generic matrices then the very argument of Lemma **1.2**, (ii) proves, relative to the basis e_{ij} for which $\xi_s = \sum x_{s,ij} e_{ij}$, $x_{s,ij}$ variables, that the elements $\xi_1^i \xi_2^j$, with $i, j = 0, \ldots, n - 1$, are a basis of $(K(x_{t,ij}))_n$ over its center $K(x_{t,ij})$. Therefore, φ satisfies the conditions of Chapter II, **5.7** (5) and we have a quotient division ring $\Lambda\langle\xi_s\rangle$ of $\Lambda\{\xi_s\}$ and inclusion maps

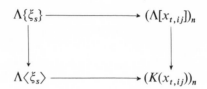

with

$$(K(x_{t,ij}))_n \simeq \Lambda\langle\xi_s\rangle \otimes_Z K(x_{t,ij})$$

(with $Z \subseteq K(x_{t,ij})$ the center of $\Lambda\langle\xi_s\rangle$). Statment (2) follows immediately.

(3)   The polynomial identities of $\Lambda\{\xi_s\}$ are the same as those of $\Lambda\langle\xi_s\rangle$ (Chapter II, **5.7**(4)) and, by what we have remarked at the beginning of the proof of (1), $\Lambda\langle\xi_s\rangle$ satisfies exactly the identities of $n \times n$ matrices.

In fact, during the proof of part (3) of the preceding theorem we have proved the following stronger statement.

Proposition 1.4   If Λ is a domain and R a Λ algebra with $\Lambda \subseteq R$, R prime of degree n, and R infinite, then R is generic in the variety of $n \times n$ matrices.

We can interpret **1.3**(3) in the sense that the variety generated by $n \times n$ matrices can be defined by equations in just two variables.

We notice finally two further corollaries.

Corollary 1.5   Let R be an infinite prime Λ algebra. Assume that $f(x)$, $g(x) \in \Lambda\{x_s\}$ are polynomials such that $f(x)g(x)$ is a polynomial identity of R. Then either $f(x)$ or $g(x)$ is a polynomial identity of R.

Proof   Let $\overline{\Lambda} = \Lambda/\mathrm{Ann}(R)$. The ring $\overline{\Lambda}$ is a domain and $\overline{\Lambda}\{\xi_s\}$, the free algebra in the variety generated by R, is a domain (**1.4**, **1.3**). Therefore, as in the map $\Pi\colon \Lambda\{x_s\} \to \overline{\Lambda}\{\xi_s\}$ we have $\Pi(f(x)g(x)) = 0$, we must have either $\Pi(f(x)) = 0$ or $\Pi(g(x)) = 0$, i.e., either $f(x)$ or $g(x)$ is a polynomial identity of R.

Corollary 1.6   Let R be an infinite prime Λ algebra. Let $f(x) \in \Lambda\{x_s\}$ be a polynomial. If $f(x)$ is *not* a polynomial identity, then $f(r)$ gives "generically" a regular element of R.

Proof   We explain in the course of the proof what we mean by "generically." First assume R is simple of dimension n^2 over its center Z; we consider R as an algebraic variety over Z. Assume $f(x_1, \ldots, x_m)$ depends on m variables; then we have a map of algebraic varieties $f: R^m \to R$ obtained by evaluation of f. Let $N: R \to Z$ be the reduced norm map and consider the composite polynomial map $Nf: R^m \to Z$. We claim that if $f \neq 0$, Nf is a nonzero map; therefore, f takes values which are invertible. To prove that Nf is a nonzero map we consider $\Lambda = \overline{\Lambda}/\mathrm{Ann}(R)$, with K its field of fractions, and the algebra $\overline{\Lambda}\{\xi_s\}$ of generic $n \times n$ matrices, free in the variety generated by R. Consider

$$
\begin{array}{ccccc}
\overline{\Lambda}\{\xi_s\} & \longrightarrow & (\overline{\Lambda}[x_{s,ij}])_n & \xrightarrow{\ \det\ } & \overline{\Lambda}[x_{s,ij}] \\
\downarrow & & \downarrow & & \downarrow \\
\overline{\Lambda}\langle\xi_s\rangle & \longrightarrow & (K(x_{s,ij}))_n & \xrightarrow{\ \det\ } & K(x_{s,ij})
\end{array}
$$

If $f \in \overline{\Lambda}\{\xi_s\}$ and $f \neq 0$, we have that f is invertible in $\overline{\Lambda}\langle\xi_s\rangle$ and therefore $0 \neq \det f \in \overline{\Lambda}[x_{s,ij}]$.

Now let $\Omega \supseteq Z$ be an algebraically closed field, $R \otimes_Z \Omega \simeq (\Omega)_n$. Choose elements $r_s \in R$; they can be thought of as matrices $r_s = \sum \alpha_{s,ij} e_{ij}$, $\alpha_{s,ij} \in \Omega$. We have a map $\psi: \overline{\Lambda}[x_{s,ij}] \to \Omega$ with $\psi[x_{s,ij}] = \alpha_{s,ij}$ and a commutative diagram

$$
\begin{array}{ccccc}
\overline{\Lambda}\{\xi_s\} & \longrightarrow & (\overline{\Lambda}[x_{s,ij}])_n & \xrightarrow{\ \det\ } & \overline{\Lambda}[x_{s,ij}] \\
\varphi\downarrow & & \psi_n\downarrow & & \psi\downarrow \\
R & \longrightarrow & (\Omega)_n & \xrightarrow{\ \det\ } & \Omega
\end{array}
$$

where $\varphi(\xi_s) = r_s$.

The bottom map is just the norm map composed with the inclusion $Z \subseteq \Omega$. Now if we had $Nf = 0$ we would have $\psi(\det f(\xi_s)) = 0$ for all $\psi: \overline{\Lambda}[x_{s,ij}] \to \Omega$ coming from elements $r_s \in R$. This therefore gives a polynomial map

$$
R^m \longrightarrow (\Omega)_n^m \xrightarrow{\ f\ } (\Omega)_n \xrightarrow{\ \det\ } \Omega
$$

which vanishes identically; the same map would vanish on $(R \otimes_Z \Omega)^m$ since Z is infinite. This is a contradiction since the polynomial $\det f(\xi_s) \neq 0$ so we can choose elements $\alpha_{s,ij} \in \Omega$ for which it does not vanish. More generally if R is an infinite prime ring and Q is its ring of quotients, R and Q have the same polynomial identities. If f is not an identity of R then on a nonempty open set U of Q^m we have that f gives invertible elements. Now R is dense in Q and so R^m is dense in Q^m; it follows that $U \cap R^m \neq \phi$ and on this set f takes elements invertible in Q and therefore regular in R. This is the exact meaning of our expression "generically."

§2   Structure of T Ideals

We consider in this paragraph an infinite field F, and $F\{x_s\}$ the free algebra in the variables x_s. If we have just one variable x it is easy to study all T ideals; they are (0) and (x^n), with $n = 1, 2, \ldots$. So we assume we have at least two variables.

It is rather clear a priori that T ideals appear in great abundance and a complete classification is almost certainly impossible; on the other hand, we will classify completely the nil radicals of all T ideals.

If I is a T ideal, either $I = (0)$, and then the nil radical of $F\{x_s\}/I = F\{x_s\}$ is (0), or $I \neq 0$, and then $F\{x_s\}/I$ is a PI algebra with its nil radical equal to the lower nil radical.

We call an ideal I of the ring R semiprime if R/I is semiprime.

Lemma 2.1   If R is a semiprime algebra over F then either R is generic in the category of all algebras or R is equivalent to $(F)_n$ for some n.

Proof   If R satisfies no polynomial identities, R is generic in the category of all algebras. Otherwise we have an embedding $R \to \oplus (\Lambda_{n_i})_{n_i}$, with $\Lambda_{n_i} \supseteq F$ a commutative algebra, and $\oplus (\Lambda_{n_i})_{n_i}$ satisfies all the stable identities satisfied by R (Chapter II, **3.1**). Now all identities of R are stable (Chapter I, **3.16**) so R and $\oplus_i(\Lambda_{n_i})_{n_i}$ are equivalent. Let $m = \max n_i$. If $k \leq m$ and Γ is a commutative F algebra, then $(\Gamma)_k$ is a specialization of $(F)_m$ and if $k = m$, then $(\Gamma)_k$ is equivalent to $(F)_m$; therefore, $\oplus_i(\Lambda_{n_i})_{n_i}$ is equivalent to $(F)_m$ and finally R is equivalent to $(F)_m$.

In general we cannot say that if R is a semiprime algebra then $f(x)g(x)$ vanishes on R if and only if one of the two is a polynomial identity of R.

This of course is the case if R is an algebra over an infinite field, by **2.1**; we have, nevertheless, the following.

Proposition 2.2   If R is a semiprime ring satisfying $S_m{}^k$ then R satisfies S_m.

Proof   We have $(0) = \{\cap P | P$ prime ideal of $R\}$ and R/P satisfies $S_m{}^k$. So we are reduced to show that $R = R/P$ satisfies S_m for all P. Now if R/P is infinite, this is given by **1.5**; otherwise R/P, being a finite prime ring, must be a matrix ring $(F)_n$ over a finite field and we only have to check that $m > 2n$. Suppose to the contrary that $m < 2n$. If $m = 2k + 1$, choose the m elements

$$e_{12}, e_{22}, e_{23}, e_{33}, \ldots, e_{kk}, e_{k\,k+1}, e_{k+1\,k+1}, e_{k+1\,1}$$

(with $k + 1 \leqslant n$ since $m + 1 = 2(k + 1) \leqslant 2n$) to substitute for the variables $x_1, \ldots, x_m$ in $S_m{}^k$. Computing S_m we see that

$$S_m(e_{12}, e_{22}, e_{23}, \ldots)$$

is a diagonal matrix

$$\begin{pmatrix} 1 & 0 & 0 & \ldots\ldots & 0 \\ 0 & * & 0 & \ldots\ldots & 0 \\ 0 & 0 & * & 0 \ldots & 0 \\ \vdots & & & & \\ 0 & 0 & \ldots\ldots\ldots & & * \end{pmatrix}$$

so $S_m{}^p \neq 0$ for every p.

If $m = 2k$ we choose instead $e_{12}, e_{22}, \ldots, e_{kk}, e_{k\,k+1}, e_{k+1\,1}$ and we get the same result.

Theorem 2.3
(1)   If I is a semiprime T ideal then either $I = (0)$ or $I = I_m =$ ideal of polynomial identities of $m \times m$ matrices.
(2)   We have $I_1 \supset I_2 \supset I_3 \supset \ldots$, all inclusions are proper.
(3) $\cap I_n = 0$.

Proof   (1)   Let $I \neq (0)$ be a semiprime T ideal. Then $F\{x_s\}/I = U$ is a semiprime PI algebra. By Lemma **2.1** the identities of U are the same as the identities of $n \times n$ matrices for some n, unless U satisfies no identities at all, i.e., $I = I_n$ or $I = (0)$. Furthermore I_n is semiprime from **1.3**.

(2)   We have often implicitly used that $I_n \supseteq I_m$ if $n \leqslant m$ since $n \times n$ matrices can be embedded (without preserving 1) in $m \times m$ matrices. Now we know that $F\{x_s\}/I_m$ is an order in a division ring of dimension m^2 over its center, by **1.3**. This is clearly an invariant of $F\{x_s\}/I_m$, hence $I_m \neq I_n$ if $m \neq n$.

(3)   Let $f(x) \in F\{x_s\}$ be a nonzero polynomial. We have to find a k such that $f(x)$ does not vanish on $k \times k$ matrices. Let s be the degree of $f(x)$ and A_{s+1} the ideal of polynomials which contain only monomials of degree $> s$. We clearly have $f(x) \notin A_{s+1}$ and so $f(x)$ is not a polynomial identity of $F\{x_s\}/A_{s+1}$. Now this last algebra is a finite dimensional algebra and therefore it can be embedded in $(F)_k$ for some k.

Remark   Part (3) had in fact already been proved in a more precise form in Chapter I, **5.1**. We have included the above proof because of its simplicity.

We are now ready to classify the nil radicals of T ideals.

Theorem 2.4   If I is a T ideal and N is its nil radical, then

(1) N is a T ideal,
(2) $N = (0)$ or $N = I_m$, $m = 1, \ldots$.

Proof   (2) is consequence of (1) and **2.1**. We prove (1). Let J be the T ideal of polynomial identities of $F\{x_s\}/N$. We have clearly $I \subseteq J \subseteq N$. Now $F\{x_s\}/N$ is semiprime and therefore by **2.1** the ideal $J = (0)$ or $J = I_m$ for some m. It follows that J is semiprime and therefore $J = N$. The theorem is fully proved.

We can interpret all these results in the language of varieties.

Definition 2.5   We call $\mathscr{V}_n$ the variety generated by $(F)_n$.

Theorem 2.6

(1) If R is a semiprime PI algebra then $R \in \mathscr{V}_n$ for some n. This n is the degree of R.

(2) If $\mathscr{V}$ is a proper variety then there is an n such that $\mathscr{V} \supseteq \mathscr{V}_n$, $\mathscr{V} \not\supseteq \mathscr{V}_{n+1}$. If R is any algebra in $\mathscr{V}$, then the degree of $R \leqslant n$.

If we do not work over an infinite field, the situation is a little different; we leave the analysis to the reader with the following exercise.

Exercise   If R is a prime Λ algebra, the ideal of identities of R is semiprime if and only if R is infinite or R is commutative.

In any case there is, a priori, a finite procedure to find the ideal of $(F)_n$ in a finite number of variables, with F a field with q elements. (It is given by Chapter I, **4.6**.) If $x_1, \ldots, x_n$ are the variables under consideration and $F[\bar{x}_{t,ij}]$ the polynomial ring over $\bar{x}_{t,ij}$, then the generic algebra associated with F is

$$\frac{F[\bar{x}_{t,ij}]}{(\bar{x}_{t,ij}^q - \bar{x}_{t,ij})} = F[\tilde{x}_{t,ij}].$$

This is a finite ring, and the matrices $\xi_t = \sum \tilde{x}_{t,ij} e_{ij}$ generate a finite algebra $F\{\xi_t\}$ which is, by Chapter I, **1.6**, isomorphic to $F\{x_s\}/I$, with I the ideal of identities of $(F)_n$.

§3   Polynomial Identities of Endomorphism Rings

We can now go back to some unfinished work done in Chapter I. We studied in Theorem **4.6** the following situation: R, S are Γ algebras, $\Gamma \supseteq \Lambda$, and S is a finite free module over Γ with basis $u_1, \ldots, u_m$; $U = \Gamma\{x_{t,i}\}$ is the free algebra in the variables $\bar{x}_{t,i}$, with $t \in T$, $i = 1, \ldots, m$, in the variety of Γ algebras generated by R. Then we proved that the free algebra in the variables x_t, for the Λ algebra $R \otimes_\Gamma S$, is isomorphic to the image of the map $\psi: \Lambda\{x_t\} \to U \otimes_\Gamma S$ sending x_t into $\sum \bar{x}_{t,i} \otimes u_i$.

We now want to deduce some specific polynomial identities for R and S through the knowledge of some suitable identities of R and S.

We know that R satisfies S_{2n}^k for suitable n, k, and also S satisfies S_{2p}^h for suitable p, h. We claim the following.

Theorem 3.1   The ring $R \otimes_\Gamma S$ satisfies $S_{2np}(x_1, \ldots, x_{2np})^j$ for some suitable j.

Proof   We take U as the free algebra in the variables $\bar{x}_{t,j}$, with $t = 1, \ldots, 2np$. Consider the nil radical $N(U)$; since $N(U)$ is locally nilpotent, the ideal $M = N(U) \otimes S$ of $U \otimes S$ is nil. Now $(U \otimes S)/M \simeq U/N(U) \otimes_\Gamma S$. The ring $U/N(U)$ can be embedded in a direct sum of matrix rings $\oplus (\Lambda_i)_{n_i}$ over commutative Γ algebras Λ_i with $n_i \leqslant n$. Therefore, since S is Γ-free, $U/N(U) \otimes_\Gamma S \subseteq \oplus (\Lambda_i \otimes_\Gamma S)_{n_i}$.

Let P_i be the nil radical of $\Lambda_i \otimes_\Gamma S$, then P_i is also locally nilpotent since $\Lambda_i \otimes S$ is a finite free module over the commutative ring Λ_i. Consider the image $\bar{S}$ of S in $(\Lambda_i \otimes_\Gamma S)/P_i$. Since the map $S \to (\Lambda_i \otimes S)/P_i$ is an extension and the last ring is semiprime, we must have that $\bar{S}$ is also semiprime (Chapter II, **6.7**). Now $\bar{S}$ satisfies S_{2p}^h and therefore it must satisfy S_{2p} (by **2.2**), so $(\Lambda_i \otimes S)/P$, which is spanned over Λ_i by $\bar{S}$, must satisfy S_{2p} (Chapter I, **3.13**). This means that $(\Lambda_i \otimes S)/P_i$ can be embedded in a direct sum of matrix rings over commutative rings of rank less than or equal to p. Finally,

$$\oplus \left(\frac{\Lambda_i \otimes_\Gamma S}{P_i} \right)_{n_i} \simeq \frac{\oplus (\Lambda \otimes S)_{n_i}}{\oplus (P_i)_{n_i}} = W$$

can be embedded in a direct sum of rings of matrices of rank less than or equal to np. This implies that W satisfies S_{2np}.

Now $\oplus (P_i)_{n_i}$ is a nil ideal, since P_i is locally nilpotent for all i; therefore, if we consider $S_{2np} \in Z\{x_1, x_2, \ldots, x_{2np}\}$ and the map

$$\psi: Z\{x_1, x_2, \ldots, x_{2np}\} \to U \otimes_\Gamma S \text{ (with } \psi(x_t) = \sum \bar{x}_{t,i} \otimes u_j),$$

we see that $\psi(S_{2np})$ vanishes under the map

$$U \otimes S \to U/N(U) \otimes S \to \oplus (\Lambda_i \otimes_\Gamma S)_{n_i} \to W.$$

The kernel of this map is clearly nil so this means that $(S_{2np})^j = 0$ for some

j. Now we know that ker ψ is exactly the ideal of polynomial identities of $R \otimes_{\Gamma} S$ so that S_{2np}^{j} is a polynomial identity of this algebra.

In particular we can apply the theorem to the following situation.

Theorem 3.2   Let R be a ring satisfying S_{2m}^{k} and let M be an R module generated by p elements. Then the ring $\mathrm{Hom}_{R}(M, M)$ satisfies S_{2mp}^{j} for some j.

Proof   Let $\pi: R^{p} \to M$ be an onto map, $K = \ker \pi$. Let

$$T = \{f \in \mathrm{Hom}_{R}(R^{p}, R^{p}) \simeq (R)_{p} | f(K) \subseteq K\} \text{ and } I$$

$$= \{f \in (R)_{p} | f(R^{p}) \subseteq K\}.$$

T is clearly a subring of $(R)_{p}$ and I an ideal of T. Furthermore, we have a map $T \to \mathrm{Hom}_{R}(M, M)$ given by factoring:

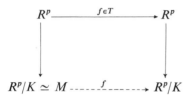

This map is onto $\mathrm{Hom}_{R}(M, M)$ since R^{p} is free and its kernel is exactly I. Therefore, $\mathrm{Hom}_{R}(M, M)$ is isomorphic to T/I. To prove the theorem it suffices, then, to show that $(R)_{p}$ satisfies S_{2mp}^{j} for some j. Now $(R)_{p} \simeq R \otimes_{Z} (Z)_{p}$, so this is a consequence of the previous theorem.

It is clear that in these lines of thought we might ask a very natural question: If R_{1}, R_{2} are two algebras satisfying, respectively, S_{2m}^{k}, S_{2p}^{h}, is it then true that $R_{1} \otimes R_{2}$ satisfies S_{2mp}^{j} for some j?

An important theorem of Regev [75] implies that $R_{1} \otimes R_{2}$ satisfies some identity and we can make predictions about its size. However, it seems reasonable to conjecture that an answer to the preceding question is "no," and in fact we conjecture that one should not be able to bound the minimum t such that $R_{1} \otimes R_{2}$ satisfies S_{t}^{j} for some j, just in terms of m, p.

Chapter IV

REPRESENTATIONS AND THEIR INVARIANTS

§1   Representations

We will use in this chapter a more categorical language, essentially the language of representable functors; this makes the treatment a little smoother.

We start with the following problem. Fix a natural number n. If S is a Λ algebra we construct $(S)_n$, the algebra of $n \times n$ matrices. This construction is clearly a functor in the category $_\Lambda \mathscr{A}$. We ask whether this functor has a left adjoint or, what is the same, whether fixed $R \in {}_\Lambda \mathscr{A}$ the functor of S: $\mathscr{M}_\Lambda(R, (S)_n)$ is representable.

Consider the algebra $(\Lambda)_n$; we have $(S)_n = S \otimes_\Lambda (\Lambda)_n$ and, therefore, a canonical map $(\Lambda)_n \xrightarrow{i} (S)_n$. Consider next the free product $R *_\Lambda (\Lambda)_n$ and the two canonical maps

$$j_1 \colon R \to R *_\Lambda (\Lambda)_n, \qquad j_2 \colon (\Lambda)_n \to R *_\Lambda (\Lambda)_n.$$

The universal property of a free product says that for any algebra W there is a 1–1 correspondence between $\mathscr{M}_\Lambda(R * (\Lambda)_n, W)$ and $\mathscr{M}_\Lambda(R, W) \times \mathscr{M}((\Lambda)_n, W)$ given by $\psi \leftrightarrow (\psi j_1, \psi j_2)$.

An elementary property that $(\Lambda)_n$ shares with any other Azumaya algebra and that we have already used in Chapter I, §9 is that any map

$(\Lambda)_n \xrightarrow{\gamma} U$ induces an isomorphism $(\Lambda)_n \otimes U^{(\Lambda)_n} \simeq U$ (where $U^{(\Lambda)_n} = \{u \in U | u\gamma(t) = \gamma(t)u$ for every $t \in (\Lambda)_n\}$).

In particular, let $C_n(R) = R * (\Lambda)_n^{(\Lambda)_n}$. We have seen that we have a canonical isomorphism

$$(C_n(R))_n \simeq C_n(R) \otimes_\Lambda (\Lambda)_n \simeq R * (\Lambda)_n.$$

If $\varphi: R \to S \otimes_\Lambda (\Lambda)_n = (S)_n$ is any morphism, then φ extends uniquely to a morphism $\tilde{\varphi}: R * (\Lambda)_n \to (S)_n$ if we require that $\tilde{\varphi}j_2$ should be the canonical map $i: (\Lambda)_n \to (S)_n$.

It is clear that the centralizer of $j_2(\Lambda)_n$ in $R * (\Lambda)_n$, i.e., $C_n(R)$, is mapped by $\tilde{\varphi}$ in the centralizer of $i(\Lambda)_n$ in $(S)_n$, i.e., in S. We have, therefore, an induced mapping $\varphi_*: C_n(R) \to S$ and it is easy to verify that we have established in this way a 1–1 correspondence between $\mathcal{M}_\Lambda(R, (S)_n)$ and $\mathcal{M}_\Lambda(C_n(R), S)$ which is natural in R and S. Therefore we have proved the following theorem.

Theorem 1.1 $S \to (S)_n$ and $R \to C_n(R)$ are adjoint functors in $_\Lambda \mathcal{A}$.

To specialize our discussion, we consider the previous situation when S is commutative (one could more generally assume that S lies in a fixed variety of algebras). It is clear that if S is commutative, $\mathcal{M}_\Lambda(C_n(R), S) = \mathcal{M}_\Lambda(C_n(R)/J, S)$ where J is the commutator ideal of $C_n(R)$. Hence the corollary.

Corollary 1.2   The functor $S \to \mathcal{M}_\Lambda(R, (S)_n)$ is representable in the category of all algebras as well as in the category of all commutative algebras (or more generally in a variety).

Consider next the canonical map

$$R \xrightarrow{j_1} R * (\Lambda)_n \simeq (C_n(R))_n.$$

It is clear, by the universal property of the construction, that the matrix elements of $j_1(r)$, with $r \in R$, generate $C_n(R)$ as a Λ algebra; therefore, we have the following proposition.

Proposition 1.3   If R is a finitely generated Λ algebra, the same is true for $C_n(R)$. More precisely, if R can be generated by k elements then $C_n(R)$ can be generated by kn^2 elements.

In particular, if $R = \Lambda\{x_1, \ldots, x_k\}$, then setting $j_1(x_i) = \sum_{s,t=1}^{n} \xi_{i,st} e_{st}$ we have $C_n(R) = \Lambda\{\xi_{i,st}\}$ and these variables are free over Λ (Chapter I, §9).

We now consider the following problem. Classify the elements in $\mathcal{M}(R, (S)_n)$, with S a commutative Λ algebra, up to the equivalence relation ρ, whereby two maps $\psi_1, \psi_2 : R \to (S)_n$ are equivalent according to ρ if there is an S automorphism γ of $(S)_n$ such that $\gamma\psi_1 = \psi_2$.

Using a geometric language, we can set the problem in the following form:

(I)   Let $V_n(R) = C_n(R)/J$ (where J is the commutator ideal), then $V_n(R)$ is a commutative Λ algebra. Consider $\mathrm{Spec}(V_n(R))$; this is an affine scheme. $\mathcal{M}_\Lambda(V_n(R), S)$ is the set of points of $\mathrm{Spec}(V_n(R))$ with coordinates in S.

(II)   Let $\mathcal{A}_n(S)$ be the group of S automorphisms of $(S)_n$. $\mathcal{A}_n(S)$ is a group-valued functor on the category of commutative Λ algebras. This functor is representable. In fact, to give an automorphism of $(S)_n$ means exactly to give a linear map $\varphi : (S)_n \to (S)_n$, invertible and compatible with the product. Thus $\varphi(e_{ij}) = \sum \eta_{ijst} e_{st}$ must be given in such a way that it should be invertible, i.e., there exists a $z \in S$ with $\det(\eta_{(ij)(st)})z = 1$. Furthermore, we have to satisfy the equation $\varphi(e_{ij})\varphi(e_{sk}) = \delta_{js}\varphi(e_{ik})$. We thus have to give a homomorphism from the algebra $\Lambda[x_{ijst}, y]$ modulo the equations $\det(x_{(ij)(st)})y - 1$ and $\sum_v x_{ijuv}x_{skvt} = \delta_{js}x_{ikut}$ into S.

Let us call $\Lambda(n)$ this algebra and note that it is finitely generated over Λ.

Proposition 1.4   The ring $\Lambda(n)$ is an Hopf algebra, i.e., $\mathrm{Spec}(\Lambda(n))$ is an affine group scheme.

Proof $\Lambda(n)$ represents a group-valued functor.

Proposition 1.5   There is a canonical action of $\mathrm{Spec}(\Lambda(n))$ on $\mathrm{Spec}(V_n(R))$.

Proof   The action $\mathscr{A}_n(S) \times \mathscr{M}_\Lambda(V_n(R), \ S) \to \mathscr{M}_\Lambda(V_n(R), \ S)$ induced by the action $\mathscr{A}_n(S) \times \mathscr{M}_\Lambda(R, \ (S)_n) \to \mathscr{M}_\Lambda(R, \ (S)_n)$ and the isomorphism $\mathscr{M}_\Lambda(V_n(R), \ S) \to \mathscr{M}_\Lambda(R, \ (S)_n)$ is natural in S; therefore, it is a natural transformation of $\mathscr{M}_\Lambda(\Lambda(n) \otimes_\Lambda V_n(R), \ S) \to \mathscr{M}_\Lambda(V_n(R), \ S)$. By Yoneda's lemma it comes from a homomorphism $V_n(R) \to \Lambda(n) \otimes V_n(R)$ [or equivalently $\mathrm{Spec}(\Lambda(n)) \times_{\mathrm{Spec}\Lambda} \mathrm{Spec}(V_n(R)) \to \mathrm{Spec}(V_n(R))$].

We indicate by $\mathscr{A}_n$ the affine group scheme $\mathrm{Spec}(\Lambda(n))$.

The equivalence classes of $\mathscr{M}_\Lambda(R, \ (S)_n)$ under the equivalence relation ρ are exactly the orbits of $\mathscr{M}_\Lambda(V_n(R), \ S)$ modulo the group $\mathscr{A}_n(S)$; we have to see whether we can construct a quotient in a suitable category. In other words, we consider the functor

$$S \to \mathscr{M}_\Lambda(R, \ (S)_n)/\rho = \mathscr{M}_\Lambda(V_n(R), \ S)/\mathscr{A}_n(S)$$

and we ask whether it is representable. The answer is no if we want to represent it in the category of rings or, dually, with a scheme which need not be affine.

To see the reason we consider a single example. Let Λ be a field, with $R = \Lambda[x]$. Then $V_n(R) = \Lambda[x_{ij}]$, the ring of polynomials in n^2 variables, and $\mathscr{M}_\Lambda(R, \ (S)_n) = (S)_n$. Let us restrict ourselves to the case where S is a field. The classes $Q_n(S) = \mathscr{M}_\Lambda(V_n(R), \ S)/\mathscr{A}_n(S)$ are exactly the conjugacy classes of matrices. Now $Q_n(S)$ cannot have a structure of variety in such a way that $(S)_n \to Q_n(S)$ is a morphism, since the group $\mathscr{A}_n(S)$ has orbits not closed in $(S)_n$ in the Zariski topology over S.

Therefore, we will have to content ourselves with a suitable open set of $\mathrm{Spec}(V_n(R))$), stable under the action of the group, over which the quotient is possible. A further remark will oblige us to slightly change our point of view.

Let B be a rank n^2 Azumaya algebra over a commutative Λ algebra A. Let $\varphi\colon R \to B$ be a morphism. Assume that $A \to S$ is a faithfully flat map of commutative algebras such that $B \otimes_A S$ is isomorphic to $(S)_n$. If we choose an isomorphism $B \otimes_A S \xrightarrow{\ \gamma\ } (S)_n$ we get, composing with φ, a map $\varphi'\colon R \to B \to B \otimes S \to (S)_n$, i.e., an element of $\mathscr{M}_\Lambda(R, \ (S)_n)$. This element is defined up to automorphisms of $(S)_n$; i.e., we have really defined an element $\bar{\varphi}$ of $\mathscr{M}_\Lambda(V_n(R), \ S)/\mathscr{A}_n(S) = \mathscr{G}_n(S)$. Consider further the two canonical maps $\varepsilon_1, \ \varepsilon_2\colon S \to S \otimes_A S$ (with $\varepsilon_1(s) = s \otimes 1$, $\varepsilon_2(s) = 1 \otimes s$). These maps induce two morphisms $\mathscr{G}_n(S) \overset{\varepsilon_{2*}}{\underset{\varepsilon_{1*}}{\rightrightarrows}} \mathscr{G}_n(S \otimes_A S)$ which clearly send the element $\bar{\varphi}$ into the same element of $\mathcal{G}_n(S \otimes_A S)$. This implies that if we had already found an open set of $\mathrm{Spec}(V_n(R))$ over which the quotient is possible—in particular if there is a subset $Q_n(S)$ of $\mathcal{G}_n(S)$ representing the points in S of this quotient and if $\bar{\varphi} \in Q_n(S)$—then, since $\bar{\varphi}$ in $Q_n(S \otimes_A S)$ is sent by the two maps ε_{1*}, ε_{2*} into the same element, $\bar{\varphi}$ must come from an element of $Q_n(A)$. Therefore, we must have a morphism $R \to (A)_n$ which, extended to $R \to (A)_n \to (S)_n$, gives exactly $\varphi': R \to B \to B \otimes S \to (S)_n$. This is clearly false for many morphisms; for instance, if $R = B$, $B \not\cong (A)_n$, and we choose the identity map for φ, it is clear that we have a counterexample.

We have thus shown that we again have to change our point of view to allow the appearence of morphisms into rank n^2 Azumaya algebras. We fix R from now on.

Definition 1.6

(1) If A is a commutative Λ algebra and B is an Azumaya algebra over A, then a map $\varphi: R \to B$ is an irreducible representation if $\varphi(R)A = B$. If B has rank n^2 over A then φ is an irreducible representation of R of degree n over A.

(2) Two irreducible representations $\varphi_1: R \to B_1$, $\varphi_2: R \to B_2$ into two Azumaya algebras over A will be called equivalent if there exists an isomorphism $\gamma: B_1 \to B_2$ of A algebras such that $\gamma\varphi_1 = \varphi_2$ (cf. Chapter II, **7.3**)

We define $Q_n(A)$ as the set of equivalence classes of irreducible representations of R of degree n over A (it is clearly a set). Q_n is easily seen to be a covariant functor in A. We wish to prove that Q_n is representable by an open set of an affine scheme; this open set will be a quotient of a suitable open set of $\mathrm{Spec}(V_n(R))$ under the action of the group $\mathcal{A}_n$.

For this purpose we make first a "local" study of our situation in the following sense. A rank n^2 Azumaya algebra B over A is not necessarily a free module over A; on the other hand, it certainly is a free module if A is local. In any case, if $u_1, \ldots, u_{n^2} \in B$ are n^2 elements, they form a basis of B over A if and only if the determinant of the matrix $\|\mathrm{Tr}(u_i u_j)\|$, formed by the reduced traces of these elements, is invertible in A; this determinant will be indicated by $\Delta(u_1, \ldots, u_n)$ and called the discriminant of the n^2-tuple.

The ideal of A generated by these discriminants is A (see [25, 40]). We now fix an n^2-tuple $\alpha = (r_1, \ldots, r_{n^2})$ of elements of R, and a morphism

$\varphi: R \to B$. If, under this morphism, the elements $\varphi(r_1), \ldots, \varphi(r_{n^2})$ form a basis of B, then certainly $\varphi(R)A = B$; therefore, φ would be an irreducible representation of degree n over A. On the other hand, we have seen that the condition that $\varphi(r_1), \ldots, \varphi(r_{n^2})$ be a basis of B is that

$$\Delta(\varphi(r_1), \ldots, \varphi(r_{n^2}))$$

be invertible in A.

Sublemma   Let S be a rank n^2 Azumaya algebra over its center A. Let $R \subset S$ be a subring containing a basis $u_1, \ldots, u_{n^2}$ of S over A. We assume that R is closed under the reduced trace map $\mathrm{Tr}: S \to A$ and that the discriminant $d = \det(\mathrm{Tr}(u_i u_j))$ of the basis u_i is invertible in R. Then R is a rank n^2 Azumaya algebra over its center B.

Proof   First we show that $u_1, \ldots, u_{n^2}$ is a basis of R over B. In fact if $r \in R$ we have $r = \sum \alpha_i u_i$, with $\alpha_i \in A$; on the other hand, the α_i's can be computed by the formulas $\mathrm{Tr}(r u_j) = \sum \alpha_i \mathrm{Tr}(u_i u_j)$. Since $d^{-1} \in R$ and R is closed under the reduced trace map, we see that $\alpha_i \in B$. This implies in particular that $S = R \otimes_B A$ and $R \to S$ is a monomorphism.

Next let us consider the following diagram:

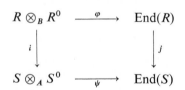

where all maps are injective and ψ is an isomorphism by hypothesis. To complete the proof of the sublemma it will be sufficient to show that φ is an isomorphism. We claim for this that $\varphi(u_i \otimes u_j)$ is a basis of $\mathrm{End}(R)$ over B. We know that this is a basis of $\mathrm{End}(S)$ over A. On the other hand, we notice that if $s, v \in S$ then $\mathrm{Tr}(\psi(s \otimes v))$ (trace in $\mathrm{End}(S)$) is just $\mathrm{Tr}(s) \cdot \mathrm{Tr}(v)$ (reduced traces). This can be proved by first splitting the Azumaya algebra S and computing the trace as an ordinary trace of matrices. Therefore, we can apply the previous reasoning to $\mathrm{End}(R) \subseteq \mathrm{End}(S)$, using the fact that $\det(\mathrm{Tr}((u_i \otimes u_j)(u_s \otimes u_t))) = d^{2n}$ is again invertible in B. This implies, as before, that $\mathrm{End}(R)$ has the elements $u_i \otimes u_j$ as a basis over B and so $\mathrm{End}(R) \simeq R \otimes R^0$.

Consider now the universal map $\lambda: R \to (V_n(R))_n$. Let $T_n(R)$ be the subalgebra of $V_n(R)$ generated by the coefficients of the characteristic polynomials of the elements $\lambda(r)$, $r \in R$. If $\alpha = (r_1, \ldots, r_{n^2})$ we indicate by d_α the element $\det(\mathrm{Tr}(\lambda(r_i r_j)))$. Clearly $d_\alpha \in T_n(R)$.

Let $Q_n(A)_\alpha$ be the subset of $Q_n(A)$ formed by all the classes of the morphisms $\varphi \colon R \to B$ for which $\varphi(r_1), \ldots, \varphi(r_{n^2})$ is a basis of B over A. We show that this functor is representable.

Consider $T_n(R)[1/d_\alpha] \subseteq V_n(R)[1/d_\alpha]$ and the subalgebra

$$W_\alpha = T_n(R)[1/d_\alpha]\lambda(R) \subseteq (V_n(R)[1/d_\alpha])_n.$$

W_α clearly has a basis over $T_n(R)[1/d_\alpha]$ formed by the elements $\lambda(r_1), \ldots, \lambda(r_{n^2})$ since, if $r \in R$, then $\lambda(r) = \sum \alpha_i \lambda(r_i)$ where the α_i can be computed by the equations $\mathrm{Tr}(\lambda(r_i r_j)) = \sum \alpha_i \mathrm{Tr}(\lambda(r_i r_j))$.

The fact that the discriminant of $\lambda(r_1), \ldots, \lambda(r_{n^2})$ is invertible in $T_n(R)[1/d_\alpha]$ and that $\mathrm{Tr} \colon W_\alpha \to T_n(R)[1/d_\alpha]$ assures that W_α is an Azumaya algebra over $T_n(R)[1/d_\alpha]$ with basis $\lambda(r_1), \ldots, \lambda(r_{n^2})$ (by the sublemma).

Lemma 1.7   The ring $T_n(R)[1/d_\alpha]$ represents the functor $Q_n(A)_\alpha$. More precisely, if $\psi \colon T_n(R)[1/d_\alpha] \to A$ is a morphism, then $W_\alpha \otimes_\psi A$ is a rank n^2 Azumaya algebra over A and $R \to W_\alpha \to W_\alpha \otimes_\psi A$ is an irreducible representation in $Q_n(A)_\alpha$. In this fashion we establish a 1–1 correspondence between $\mathscr{M}(T_n(R)[1/d_\alpha], A)$ and $Q_n(A)_\alpha$.

Proof   It is clear that we have defined a transformation

$$\eta \colon \mathscr{M}(T_n(R)[1/d_\alpha], A) \to Q_n(A)_\alpha$$

in the lemma and that it is natural. We have to prove that it is bijective.

Therefore, let $R \xrightarrow{\varphi} B$ be an element whose equivalence class is in $Q_n(A)_\alpha$. Let $A \to S$ be a faithfully flat morphism splitting B, i.e., $B \otimes_A S$ is isomorphic to $(S)_n$. We fix a particular isomorphism $B \otimes_A S \xrightarrow{\gamma} (S)_n$. From the universal properties of the map $\lambda \colon R \to (V_n(R))_n$ we can construct a map $\varphi_* \colon V_n(R) \to S$ such that the diagram

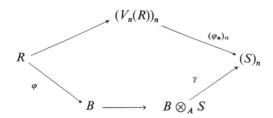

is commutative. Now $\varphi_*(d_\alpha)$ is invertible in S, since the elements $\varphi(r_i)$ form a basis of B, therefore, this diagram extends canonically as follows:

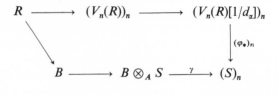

We obtain in particular, restricted to W_α, a map $\tilde{\varphi}: W_\alpha \to (S)_n$ such that the diagram

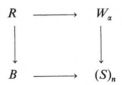

is commutative.

Now $W_\alpha = \lambda(R)T_n(R)[1/d_\alpha]$, and $T_n(R)$ is generated by the coefficients of the characteristic polynomials of the elements $\lambda(r)$, $r \in R$. Under the induced morphism $T_n(R)[1/d_\alpha] \to S$, the coefficients of the characteristic polynomial of $\lambda(r)$ are sent into the coefficients of the characteristic polynomial of $\varphi(r)$; therefore, they are elements of A. The map $\tilde{\varphi}: T_n(R)[1/d_\alpha] \to A$ is, by the same remark, independent of the particular splitting that is chosen. It follows that we can construct the canonical commutative diagram

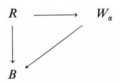

from which we deduce $B \simeq W_\alpha \otimes_{\tilde{\varphi}} A$ and $\varphi: R \to B$ is obtained through $R \to W_\alpha \to W_\alpha \otimes_{\tilde{\varphi}} A$.

This construction gives rise to a natural transformation

$$\zeta: Q_n(A)_\alpha \to \mathscr{M}(T_n(R)[1/d_\alpha], A)$$

such that $\eta\zeta$ is the identity of $Q_n(A)_\alpha$. Therefore, we have only to prove that $\zeta\eta$ is the identity of $\mathscr{M}(T_n(R)[1/d_\alpha], A)$. This last fact is actually trivial since, given $\varphi: T_n(R)[1/d_\alpha] \to A$, we consider the Azumaya algebra $W_\alpha \otimes_\varphi A$ and the morphism $W_\alpha \to W_\alpha \otimes_\varphi A$. This morphism is compatible with taking characteristic polynomials. If we split $W_\alpha \otimes_\varphi A$ and construct the induced morphism $T_n(R)[1/d_\alpha] \to A$ according to the construction of ζ, we easily see that we get back the given morphism φ.

We now have to globalize our results, which is not too hard owing to the properties of Azumaya algebras.

Let U be the open set of $\mathrm{Spec}(T_n(R))$ which is the union of all the open sets $U_\alpha = \mathrm{Spec}(T_n(R)[1/d_\alpha]) \subseteq \mathrm{Spec}(T_n(R))$. We first show that U has a coherent sheaf of Azumaya algebras. In fact, over U_α we have the Azumaya algebra W_α and it is clear that these algebras glue together in the intersections to give the desired sheaf $\mathscr{F}$ over U.

If $\Gamma(\mathscr{F})$ is the algebra of sections of $\mathscr{F}$ over U we clearly have a morphism $\lambda(R)T_n(R) \to \Gamma(\mathscr{F})$.

Now we come to the main property; we want to show that the morphisms $\mathrm{Spec}(A) \to U$ are in 1–1 correspondence with $Q_n(A)$.

If $\mu: \mathrm{Spec}(A) \to U$ is a morphism then $\mu^*(\mathscr{F})$ is a coherent sheaf of Azumaya algebras over $\mathrm{Spec}(A)$ and therefore it is the sheaf associated to an Azumaya algebra B over A.

By the construction of B we have a map $\Gamma(\mathscr{F}) \to B$ and this map, composed with $R \to \lambda(R)T_n(R) \to \Gamma(\mathscr{F})$, gives rise to a representation $\psi: R \to B$. The hypothesis that this map comes from a morphism $\mu: \mathrm{Spec}(A) \to U$ implies that we have a ring homomorphism $\psi: T_n(R) \to A$ and that the elements $\psi(d_\alpha)$ generate the unit ideal of A. Therefore $\psi(R)A \subseteq B$ is such that localization gives $\psi(R)A[1/\psi(d_\alpha)] = B[1/\psi(d_\alpha)]$. Since the $\psi(d_\alpha)$ generate A we see that $\psi(R)A = B$.

Conversely, assume that $\varphi: R \to B$ is an irreducible representation of degree n over A. If $\mathscr{M}$ is a maximal ideal of A then $B_\mathscr{M}$ is an Azumaya algebra over the local ring $A_\mathscr{M}$; in particular it is a free $A_\mathscr{M}$ module. Since $\varphi(R)A = B$ we have $\varphi(R)A_\mathscr{M} = B_\mathscr{M}$ and, $A_\mathscr{M}$ being local, we can find $\alpha = (r_1, \ldots, r_{n^2})$, with $r_i \in R$, such that the elements $\varphi(r_1), \ldots, \varphi(r_{n^2})$ are a basis of $B_\mathscr{M}$ over $A_\mathscr{M}$. If d_α is the discriminant of this basis, these elements form a basis of $B[1/d_\alpha]$ and $\varphi_\alpha: R \xrightarrow{\varphi} B \to B[1/d_\alpha]$ induces a map $\bar{\varphi}_\alpha: W_\alpha \to B[1/d_\alpha]$ making the following diagram commutative:

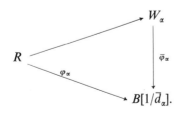

Now since $\mathcal{M}$ is any maximal ideal it is clear that the $\bar{d}_\alpha$ of this form generates the unit ideal of A. The morphism $T_n(R)[1/d_\alpha] \to B[1/\bar{d}_\alpha]$ can be interpreted dually as giving morphisms of the open sets $V_\alpha = \mathrm{Spec}(A[1/\bar{d}_\alpha])$ $\subseteq \mathrm{Spec}(A)$ in the scheme $U = \bigcup U_\alpha$ already defined. These morphisms glue together by construction and since the V_α cover $\mathrm{Spec}(A)$ we have the required morphism $\mathrm{Spec}(A) \to U$.

It is clear, localizing, that the two constructions are mutually inverse. Therefore U represents, in the category of schemes, the functor $Q_n(A)$. We have proved the following theorem.

Theorem 1.8   The functor $\mathrm{Spec}(A) \to Q_n(A)$ is represented in the category of schemes by the open set $U \subseteq \mathrm{Spec}(T_n(R))$ defined previously.

Before finishing this argument we remark that $\mathcal{M}(R, (A)_n)$ is clearly also a functor in the variable R, therefore $C_n(R)$ and $V_n(R)$ are functors of R. As for $Q_n(A)$ this is no longer true. It is clear that if $S \xrightarrow{\psi} R$ is a ring homomorphism and $\varphi: R \to B$ is in $Q_n(R)$, it is no longer necessarily true that $\varphi\psi: S \to B$ is in $Q_n(S)$ because the condition $\varphi\psi(S)A = B$ need not be satisfied.

This is the case for a particular class of maps, in fact for central extensions (Chapter II, **6.3**). We notice the following.

(1)   If we restrict our morphisms between algebras to be central extensions we have a category.

(2)   If B is an Azumaya algebra, a map $R \to B$ is an irreducible representation if and only if it is a central extension.

We see, therefore, the following corollary of the preceding theorem.

Corollary 1.9   If we write $Q_n(R, A)$ instead of $Q_n(A)$ to put in evidence the dependence on R, and similarly if we write U_R for the scheme representing $Q_n(R, A)$, then:

(1) $Q_n(R, A)$ is a set-valued functor in the category of rings whose morphisms are central extensions; and

(2) $R \to U_R$ is a functor from the category of rings whose morphisms are central extensions to the category of schemes.

§2   Invariants

The functor $Q_n(A)$ that we have constructed in the previous paragraph has been obtained, modifying in a suitable form the functor $\mathcal{M}_\Lambda(V_n(R), S)/\mathcal{A}_n(S)$, which gives the orbit space of the S points $\mathcal{M}(R, (S)_n)$ of the variety $\text{Spec}(V_n(R))$ modulo the action of the group $\mathcal{A}_n(S)$. If we consider $\alpha = (r_1, \ldots, r_{n^2})$, with $r_i \in R$, and the subset $\mathcal{M}(R, (S)_n)_\alpha$ of $\mathcal{M}(R, (S)_n)$ formed by those morphisms whose equivalence classes are in $Q_n(S)_\alpha$ we clearly see the following.

(i)   The functor $\mathcal{M}(R, (S)_n)_\alpha$ is represented by $V_n(R)[1/d_\alpha]$.

(ii) $\mathcal{M}(R, (S)_n)_\alpha$ is stable under the action of $\mathcal{A}_n(S)$.

(iii)   Globalizing and setting $\tilde{\mathcal{M}}(R, (S)_n) = \bigcup_\alpha \mathcal{M}(R, (S)_n)_\alpha$, we obtain $\tilde{\mathcal{M}}(R, (S)_n)$ represented in the category of schemes by the open set V of $\text{Spec}(V_n(R))$, where $V = \bigcup \text{Spec}(V_n(R)[1/d_\alpha])$.

(iv) V is stable under the action of the group $\mathcal{A}_n$.

(v)   There is a map $\pi: V \to U$ making the following diagram commutative:

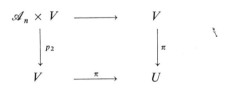

(vi)   For the geometric points in S we see that the fibers of

$$\tilde{\mathcal{M}}(R, (S)_n) \to Q_n(S)$$

are the orbits under $\mathcal{A}_n(S)$.

(vii)   If S is an algebraically closed field then any rank n^2 Azumaya algebra over S is isomorphic to $(S)_n$ so $Q_n(S)$ is exactly the orbit space $\tilde{\mathcal{M}}(R, (S)_n)/\mathcal{A}_n(S)$.

These remarks show that U is a quotient of V in some sense. We want to show that it is a geometric quotient (cf. [64]). This is now essentially a local problem and it will be enough to show for every α that on the open set U the ring of functions $T_n(R)[1/d_\alpha]$ is the invariant ring of $V_n(R)[1/d_\alpha]$ under the action of the group in the sense that we will indicate in the following analysis.

Consider $\mathscr{A}_n(\Lambda)$; if A is a commutative Λ algebra, we have a canonical morphism $\mathscr{A}_n(\Lambda) \to \mathscr{A}_n(A)$; if $g \in \mathscr{A}_n(\Lambda)$ we indicate g_A to be the corresponding element of $\mathscr{A}_n(A)$.

Lemma 2.1   If $h: A \to B$ is a morphism of commutative Λ algebras and $g \in \mathscr{A}_n(\Lambda)$, then the following diagram is commutative:

$$
\begin{array}{ccc}
(A)_n & \xrightarrow{\ g_A\ } & (A)_n \\
\downarrow{\scriptstyle h_n} & & \downarrow{\scriptstyle h_n} \\
(B)_n & \xrightarrow{\ g_B\ } & (B)_n
\end{array}
$$

Proof   The two morphisms $g_B h_n$ and $h_n g_A$ restricted to the center A of $(A)_n$ induce the same map $h: A \to B$ (with B the center of $(B)_n$). On the other hand, if we compose these maps with the canonical map $(\Lambda)_n \to (A)_n$, they clearly induce the same morphism, i.e., $(\Lambda)_n \xrightarrow{\ g\ } (\Lambda)_n \to (B)_n$. Now $(A)_n$ is generated by A and the image of $(\Lambda)_n$; therefore, the lemma is proved.

Corollary 2.2

(1) The map $\mathscr{M}(R, (A)_n) \xrightarrow{\ g_{A^*}\ } \mathscr{M}(R, (A)_n)$ given by $g_{A^*}(\varphi) = g_A\varphi$ is a natural transformation of the functor.

(2) The group $\mathscr{A}_n(\Lambda)$ operates naturally on $V_n(R)$, i.e., if $g \in \mathscr{A}_n(\Lambda)$ and $g_{A^*}: V_n(R) \to V_n(R)$ is the induced automorphism, then, for every commutative algebra A and morphism $h: V_n(R) \to A$ classifying a morphism $\varphi: R \to (A)_n$, the following diagram is commutative:

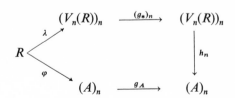

Proof (1) This is a consequence of **2.1**. (2) This follows from (1) via Yoneda's lemma.

Now we show that $T_n(R) \subseteq V_n(R)$ is invariant under the action of g because the diagram

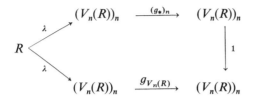

is commutative; but if $r \in R$, we have that $g_{V_n(R)}\lambda(r)$ has the same characteristic polynomial as $\lambda(r)$, since $g_{V_n(R)}$ is a $V_n(R)$ automorphism. On the other hand, $g_{V_n(R)}\lambda(r) = (g_*)_n(\lambda(r))$ and the characteristic polynomial of $(g_*)_n(\lambda(r))$ is the image via g_* of the characteristic polynomial of $\lambda(r)$. Therefore, g_* leaves invariant the characteristic polynomial of $\lambda(r)$, with $r \in R$. It follows that $T_n(R)$ is invariant.

We would like to show that $T_n(R)$ is exactly the algebra of invariants of $V_n(R)$ under the action of this group. Clearly this will be false as stated, for instance, if Λ is a finite field. We do not want the invariants just under $\mathscr{A}_n(\Lambda)$ but also under extensions of the base ring. This is not too serious a difficulty. What is really an obstacle is the fact that we have a quotient only on an open set of $\mathrm{Spec}(V_n(R))$; therefore, only on this open set can we prove a theorem on the invariant functions. We now proceed in this direction.

Let α be fixed, and consider $V_n(R)[1/d_\alpha] = V_\alpha$ and $T_n(R)[1/d_\alpha] = T_\alpha$. The element $d_\alpha \in T_\alpha$ is invariant under the action of $\mathscr{A}_n(\Lambda)$; therefore, this action can be extended to V_α, and T_α is contained in the ring of invariants.

If $\Lambda \to \Gamma$ is any map of commutative rings, we set $R_\Gamma = R \otimes_\Lambda \Gamma$ and consider R_Γ as a Γ algebra. If $\varphi: R \to S$ is a morphism of Λ algebras, we set $\varphi_\Gamma: R_\Gamma \to S_\Gamma$, the extended morphism of Γ algebras.

For R_Γ we can repeat all our constructions, working now on the category of Γ algebras. Fortunately there is a strict relationship between the old and the new constructions. In effect, if S is a Γ algebra, then S can be considered as a Λ algebra and we have

$$\mathscr{M}_\Gamma(R_\Gamma, S) \simeq \mathscr{M}_\Lambda(R, S).$$

It follows immediately that

$$V_n(R_\Gamma) = V_n(R) \otimes_\Lambda \Gamma = V_n(R)_\Gamma$$

with the canonical morphism

$$\lambda_\Gamma \colon R_\Gamma \to [(V_n(R))_n]_\Gamma = (V_n(R)_\Gamma)_n.$$

Furthermore we have the action of $\mathscr{A}_n(\Lambda)$ on $V_n(R)$ and of $\mathscr{A}_n(\Gamma)$ on $V_n(R)_\Gamma$. It is clear that these two actions are compatible in the sense that if $g \in \mathscr{A}_n(\Lambda)$ and $g_* \colon V_n(R) \to V_n(R)$ is the induced morphism, then $(g_*)_\Gamma \colon V_n(R)_\Gamma \to V_n(R)_\Gamma$ coincides with $(g_\Gamma)_*$ where g_Γ is the automorphism corresponding to g in the map $\mathscr{A}_n(\Lambda) \to \mathscr{A}_n(\Gamma)$.

Definition 2.3   We set $\mathscr{J} = \{a \in V_n(R)|$ for every morphism $\Lambda \to \Gamma$ of commutative algebras the element $a \otimes 1 \in V_n(R)_\Gamma$ is invariant with respect to $\mathscr{A}_n(\Gamma)\}$. Then $\mathscr{J}$ will be called the ring of absolute invariants of degree n irreducible representations.

Localizing, we have a similar definition of the ring $\mathscr{J}_\alpha \subseteq V_n(R)[1/d_\alpha] = V_\alpha$.

Lemma 2.4   We have $T_n(R) \subseteq \mathscr{J}$, $T_\alpha \subseteq \mathscr{J}_\alpha$.

Proof   The lemma follows from the fact that $T_n(R)_\Gamma \subseteq T_n(R_\Gamma) \subseteq$ {ring of invariants of $V_n(R_\Gamma)$ with respect to $\mathscr{A}_n(\Gamma)\}$.

Theorem 2.5   The main theorem of this section is $T_\alpha = \mathscr{J}_\alpha$.

Proof   Let $j \colon T_\alpha \to \Gamma$ be any map, with Γ a commutative algebra, such that $W_\alpha \otimes_j \Gamma$ is isomorphic to $(\Gamma)_n$ (W_α as defined in §1). We choose a particular isomorphism $\mu \colon W_\alpha \otimes_j \Gamma \to (\Gamma)_n$. The composed map

$$\psi_\mu \colon R \to W_\alpha \to W_\alpha \otimes_j \Gamma \to (\Gamma)_n$$

has a classifying map $\bar{\psi}_\mu \colon V_\alpha \to \Gamma$ and, therefore, by extending the ring, a map $V_\alpha \otimes \Gamma \to \Gamma$, i.e., a map $(V_\Gamma)_\alpha \to \Gamma$ classifying the extended map

$\bar{\psi}: R_\Gamma \to (\Gamma)_n$. This morphism clearly depends on the particular μ that we have chosen. If we change μ through an automorphism $g \in \mathscr{A}_n(\Gamma)$ we have the following commutative diagram:

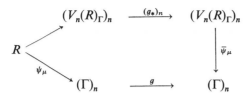

It follows that, on the invariant ring of $V_n(R)_\Gamma$ with respect to $\mathscr{A}_n(\Gamma)$, the morphism $\bar{\psi}_\mu$ is independent of the chosen μ. Therefore, j induces a well-defined map $\bar{j}: \mathscr{J}_\alpha \to \Gamma$. If we choose j faithfully flat, we construct the two canonical maps

$$\varepsilon_1, \ \varepsilon_2: \Gamma \rightrightarrows \Gamma \otimes_{T_\alpha} \Gamma \quad \text{given by} \quad \varepsilon_1(\gamma) = 1 \otimes \gamma, \quad \varepsilon_2(\gamma) = \gamma \otimes 1,$$

and the sequence of rings $T_\alpha \to \Gamma \rightrightarrows \Gamma \otimes_{T_\alpha} \Gamma$ is exact in the sense that $T_\alpha \xrightarrow{j} \Gamma$ is the equalizer of $\varepsilon_1, \varepsilon_2$ (i.e., $\varepsilon_1(\gamma) = \varepsilon_2(\gamma)$ implies that there is a unique $t \in T_\alpha$ with $\gamma = j(t)$).

Keeping the isomorphism $\mu: W_\alpha \otimes_j \Gamma \simeq (\Gamma)_n$ fixed, we consider the two isomorphisms $W_\alpha \otimes (\Gamma \otimes_{T_\alpha} \Gamma) \underset{\mu_2}{\overset{\mu_1}{\rightrightarrows}} (\Gamma \otimes_{T_\alpha} \Gamma)_n$ deduced from μ when ε_1 and ε_2 are applied. From the previous discussion, we have that $\bar{j}: \mathscr{J}_\alpha \to \Gamma$ composed with ε_1 and ε_2, respectively, is the unique map $\mathscr{J}_\alpha \to \Gamma \otimes \Gamma$ deduced from the existence of an isomorphism of $W_\alpha \otimes (\Gamma \otimes_{T_\alpha} \Gamma)$ with $(\Gamma \otimes_{T_\alpha} \Gamma)_n$, in particular $\varepsilon_1 \bar{j} = \varepsilon_2 \bar{j}$. Therefore, we can complete in a unique way the following diagram:

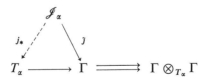

Now consider $W_\alpha \subseteq (V_\alpha)_n$. This inclusion tells us that the inclusion map $T_\alpha \to V_\alpha$ induces a splitting $W_\alpha \otimes V_\alpha \simeq (V_\alpha)_n$. We can compare this splitting with the one given by Γ; to do this we form the tensor product $(W_\alpha \otimes_{T_\alpha} V_\alpha) \otimes_{T_\alpha} \Gamma$. We have two different splittings of this algebra, the first constructed using the isomorphism $(W_\alpha \otimes V_\alpha) \simeq (V_\alpha)_n$, the second using the isomorphism $(W_\alpha \otimes \Gamma) \simeq (\Gamma)_n$. From what was remarked in the beginning of the proof, the two splittings induce the same map: $\mathscr{I}_\alpha \to V_\alpha \otimes_{T_\alpha} \Gamma$. This map, computed explicitly, is for one splitting the inclusion $\mathscr{I}_\alpha \to V_\alpha$ followed by the map $V_\alpha \to V_\alpha \otimes_{T_\alpha} \Gamma$; for the second splitting it is $j_* \colon \mathscr{I}_\alpha \to T_\alpha$ followed by the map $T_\alpha \to V_\alpha \otimes_{T_\alpha} \Gamma$.

It follows that if $u \in \mathscr{I}_\alpha$ then $u \otimes 1 = 1 \otimes j_*(u) = j_*(u) \otimes 1$. Since $T_\alpha \to \Gamma$ is faithfully flat, it follows that $u = j_*(u)$. Therefore, $\mathscr{I}_\alpha = T_\alpha$ and j_* is the identity map.

Remark   Whether the same theorem is true for V, T, and $\mathscr{I}$ is not at all clear and we leave it as an open question.

§3   Relative versus Absolute Invariants

We want to complete the discussion of the previous section and show that in some cases the ring of absolute invariants coincides with the ring of invariants with respect to the group $\mathscr{A}_n(\Lambda)$.

We recall that the action of $\mathscr{A}_n(A)$ on $\mathscr{M}(R, (A)_n)$ is natural in A and therefore it is given, by Yoneda's lemma, by a morphism:

$$V_n(R) \xrightarrow{\mu} V_n(R) \otimes_\Lambda \Lambda(n).$$

Lemma 3.1   We claim that $\mathscr{I} = \{a \in V_n(R) | \mu(a) = a \otimes 1\}$.

Proof   This is almost trivial; note that $\Gamma(n) = \Lambda(n) \otimes \Gamma$ and the action of $\mathscr{A}_n(A)$ over $\mathscr{M}_\Gamma(R_\Gamma, (A)_n)$ is classified in the category of Γ algebras by the morphism

$$V_n(R)_\Gamma \xrightarrow{\mu_\Gamma} (V_n(R) \otimes_\Lambda \Lambda(n)) \simeq V_n(R) \otimes_\Gamma \Gamma(n).$$

On the other hand, if an element u is in $\mathscr{I}$, it follows that for every $g \in \mathscr{A}_n(\Gamma)$ we have $g_*(u \otimes 1) = u \otimes 1$ in $V_n(R) \otimes \Gamma$. Interpreting everything in terms of the classifying objects and maps, we have that g_* is classified by a map $u_g \colon \Lambda(n) \to \Gamma$, i.e., $g_* \colon V_n(R)_\Gamma \to V_n(R)_\Gamma$ is just the composition $V_n(R)_\Gamma \xrightarrow{\mu_\Gamma} V_n(R)_\Gamma \otimes_\Gamma \Gamma(n) \to (V_n(R)_\Gamma \otimes \Gamma) \otimes_\Lambda \Lambda(n) \xrightarrow{u_g} V_n(R)_\Gamma \otimes_\Gamma \Gamma$, where $\bar{u}_g(b \otimes c) = b u_g(c)$.

If we restrict our computation to $V_n(R)$ we have the composition

$$V_n(R) \xrightarrow{\mu} V_n(R) \otimes \Lambda(n) \to V_n(R) \otimes \Gamma.$$

Therefore, a is an absolute invariant, i.e., $a \in \mathcal{J}$ if and only if

$$(1 \otimes u_g)\mu(a) = a \otimes 1$$

for every Γ and every $g \in \mathcal{A}_n(\Gamma)$. In particular this must happen for $\Gamma = \Lambda(n)$, and $g \in \mathcal{A}_n(\Lambda(n))$ classified by the identity map $1_{\Lambda(n)}$. Therefore we must have $\mu(a) = (1 \otimes 1)\mu(a) = a \otimes 1$. Conversely, if $\mu(a) = a \otimes 1$ we have $(1 \otimes u_g)\mu(a) = (1 \otimes u_g)(a \otimes 1) = a \otimes 1$ and so $a \in \mathcal{J}$.

We can say shortly that $1_{\Lambda(n)}$ classifies a generic automorphism and $a \in V_n(R)$ is an absolute invariant if and only if it is an invariant of this generic automorphism.

The same line of reasoning applies to V_α, in fact $d_\alpha \in \mathcal{J}$ and therefore $\mu(d_\alpha) = d_\alpha \otimes 1$; from this we see that $V_\alpha \otimes \Lambda(n)$ is obtained by inverting $d_\alpha \otimes 1$ and the deduced morphism $\mu_\alpha : V_\alpha \to V_\alpha \otimes \Lambda(n)$ is obtained simply by extending μ to this localization.

Theorem 3.2   We now see the required criterion: $\mathcal{J}_\alpha = V_\alpha^{\mathcal{A}_n(\Lambda)}$ (i.e., the invariants of $\mathcal{A}_n(\Lambda)$ are absolute invariants) if Λ is an infinite domain and V_α is without Λ torsion. (In particular these two conditions are automatically satisfied if Λ is an infinite field.)

Proof   Clearly $\mathcal{J}_\alpha \subseteq V_\alpha^{\mathcal{A}_n(\Lambda)}$; the point of the theorem is the converse. First we claim that $u \in \mathcal{J}_\alpha$ if and only if, for every commutative local algebra D, we have that $u \otimes 1 \in V_\alpha \otimes D$ is an invariant of $\mathcal{A}_n(D)$. In fact if $u \in \mathcal{J}_\alpha$ then the previous condition is true for every commutative algebra. Conversely, assume that the previous statement is true. Let Γ be any commutative algebra and $g \in \mathcal{A}_n(\Gamma)$. Let $g_* : V_\alpha \otimes \Gamma \to V_\alpha \otimes \Gamma$ be the induced map. Localizing at every maximal ideal $\mathcal{M}$ of Γ we see that

$$(g_\mathcal{M})_* : V_\alpha \otimes \Gamma_\mathcal{M} \to V_\alpha \otimes \Gamma_\mathcal{M}$$

and $(g_\mathcal{M})_* = (g_*)_\mathcal{M}$ (where $g_\mathcal{M}$ is the automorphism induced by g in $\mathcal{A}_n(\Gamma_\mathcal{M})$). If u is invariant for every local algebra, we have $(g_\mathcal{M})_*(u \otimes 1) = u \otimes 1$ for every $\mathcal{M}$; therefore, by the localization principle, $g_*(u \otimes 1) = u \otimes 1$ and $u \otimes 1$ is invariant with respect to $\mathcal{A}_n(\Gamma)$.

Now if Γ is any local ring we have the following facts:

(i)   Every automorphism of $(\Gamma)_n$ is inner.

(ii)   Every element $A \in GL(n, \Gamma)$ is equivalent, modulo the subgroup $E_n(\Gamma)$ generated by the elementary matrices $1 + \beta e_{ij}$ (with $i \neq j$), to a diagonal matrix $\mathrm{diag}(1, 1, \ldots, 1, d)$ with $d = \det A$.

(iii)   For every commutative ring, if there is a t such that $d = t^n$ then $\mathrm{diag}(1, 1, \ldots, 1, d)$ (with d invertible) is equivalent, modulo $E_n(\Gamma)$, to $\mathrm{diag}(t, t, \ldots, t) = t \cdot 1_{(\Gamma)_n}$.

Let now $u \in V^{\mathscr{A}_n(\Lambda)}$; we want to show that, given any local algebra and an element $g \in \mathscr{A}_n(\Gamma)$, we have $g_*(u \otimes 1) = u \otimes 1$ in $V_\alpha \otimes \Gamma$. This will show that $u \in \mathscr{I}_\alpha$ from the previous remarks. We know that g is an inner automorphism; let $A \in GL(n, \Gamma)$ be a matrix such that $g(B) = ABA^{-1}$ for every $B \in (\Gamma)_n$. Let $d = \det A$ and consider the algebra

$$\Gamma' = \frac{\Gamma[x]}{(x^n - d)}.$$

The ring Γ' is free over Γ; furthermore, setting $t = \bar{x}$ we have $t^n = d$ in Γ'. Thus the matrix $\mathrm{diag}(1, 1, \ldots, 1, d)$ is equivalent, modulo $E_n(\Gamma')$, to $\mathrm{diag}(t, t, \ldots, t)$. It follows that, if we extend $g \in \mathscr{A}_n(\Gamma)$ to $\mathscr{A}_n(\Gamma')$, we obtain an inner automorphism, given conjugating by the matrix A, which is in Γ' equivalent to $\mathrm{diag}(t, t, \ldots, t)$ modulo elementary matrices. Since the inner automorphism induced by a matrix does not change if we multiply the matrix by a constant diagonal matrix, we deduce that $g_{\Gamma'}$ (the automorphism induced by g in $\mathscr{A}_n(\Gamma')$) is the inner automorphism obtained by conjugating with a matrix $A' \in E_n(\Gamma')$. Now A' is a product (in the given order) $A' = \prod_{i=1}^m (1 + \alpha_i e_{s_i t_i})$, $\alpha_i \in \Gamma'$.

Let $\Lambda[x_1, \ldots, x_m]$ be the polynomial ring in m variables over Λ; consider the morphism $\pi: \Lambda[x_1, \ldots, x_m] \to \Gamma'$ given by $\pi(x_i) = \alpha_i$. The matrix $C = \prod_{i=1}^m (1 + x_i e_{s_i t_i})$ is in $E_n(\Lambda[x_1, \ldots, x_m])$ and it is mapped, in the induced morphism

$$\pi_*: GL(n, \Lambda[x_1, \ldots, x_m]) \to GL(n, \Gamma')$$

to the matrix A'.

The inner automorphism g' of $(\Lambda[x_1, \ldots, x_m])_n$ induced by C is mapped in the induced morphism $\mathscr{A}_n(\Lambda[x_1, \ldots, x_m]) \xrightarrow{\pi_*} \mathscr{A}_n(\Gamma')$, into the automorphism $g_{\Gamma'}$. Therefore, if we can show that $u \otimes 1 \in V_\alpha \otimes \Lambda[x_1, \ldots, x_m]$

is invariant under $(g')_*$ (deduced from g'), then we will have that $u \otimes 1 \in V_\alpha \otimes \Gamma'$ is invariant under the automorphism induced by $g_{\Gamma'}$. Since Γ' is free over Γ, we will finally see that $u \otimes 1 \in V_\alpha \otimes \Gamma$ is invariant with respect to g. We have to prove that $u \otimes 1 \in V_\alpha \otimes \Lambda[x_1, \ldots, x_m]$ is invariant with respect to $(g')_*$.

Let $v = (g')_*(u \otimes 1) - u \otimes 1$; we must show that $v = 0$. If $h: \Lambda[x_1, \ldots, x_m] \to \Lambda$ is any map, then h gives rise to a map

$$\mathscr{A}_n(\Lambda[x_1, \ldots, x_m]) \to \mathscr{A}_n(\Lambda).$$

If $\bar{g}'$ is the automorphism induced in this way by g', we will have $(\bar{g}')_*(u) = \tilde{h}((g')_*(u \otimes 1))$ where $\tilde{h}: V_\alpha \otimes \Lambda[x_1, \ldots, x_m] \to V_\alpha \otimes \Lambda$ is given by $\tilde{h}(a \otimes b) = ah(b)$. Since by assumption $u \in V_\alpha^{\mathscr{A}_n(\Lambda)}$, we have $(\bar{g}')_*(u) = u$ and therefore $h(v) = 0$. To resume, we have that

$$v \in V_\alpha \otimes \Lambda[x_1, \ldots, x_m] = V_\alpha[x_1, \ldots, x_m]$$

is a polynomial that, whenever computed in Λ, gives values identically 0. We have to deduce from this, under the particular conditions stated in the theorem for Λ and V_α, that v is effectively the 0 polynomial. This will finish the proof of our theorem.

Thus we have to show that if Λ is an infinite domain, Γ a Λ algebra without torsion, and $w \in \Gamma[x_1, \ldots, x_m]$ a polynomial vanishing whenever computed in Λ, then $w = 0$. We will show this by induction.

Let $m = 1$, $w = \sum_{i=0}^{n} \gamma_i x^i$. If $\lambda_1, \ldots, \lambda_{n+1} \in \Lambda$ are $n+1$ different elements, we have $\sum_{i=0}^{n} \gamma_i \lambda_j{}^i = 0$ with $j = 1, \ldots, n+1$. On the other hand, the determinant

$$d = \begin{vmatrix} \lambda_1^n & \lambda_1^{n-1} & \cdots & \lambda_1^0 \\ \lambda_2^n & \lambda_2^{n-1} & \cdots & \lambda_2^0 \\ & & \vdots & \\ \lambda_{n+1}^n & \lambda_{n+1}^{n-1} & \cdots & \lambda_{n+1}^0 \end{vmatrix} = \pm \prod_{i>j} (\lambda_i - \lambda_j) \neq 0$$

and from the previous relations, $d\gamma_i = 0$ for every i. Since Γ is without torsion we must have $\gamma_i = 0$ and so $w = 0$. In general let

$$w = \sum_{i=0}^{n} a_i(x_1, \ldots, x_{m-1}) x_m{}^i.$$

If $\lambda_1, \ldots, \lambda_{m-1}$ are arbitrary, the polynomial $\sum_{i=0}^{n} a_i(\lambda_1, \ldots, \lambda_{m-1}) x_m^i$

vanishes over Λ; from the one variable case, it follows that

$$a_i(\lambda_1, \ldots, \lambda_{m-1}) = 0.$$

This is true for all choices $\lambda_1, \ldots, \lambda_{m-1} \in \Lambda$ and so, by induction, the polynomials $a_i(x_1, \ldots, x_{m-1})$ are identically zero. This finally implies that $w = 0$.

Remark   We have seen, during the proof of the theorem, that the only necessary hypothesis is that a polynomial $w \in V_\alpha[x_1, \ldots, x_m]$ vanishing on Λ is necessarily 0. Therefore the theorem holds in this generality.

We also remark that the theorem proved is really a theorem on the group $\mathscr{A}_n$ and not on the particular action considered.

§4   Finite Generation of the Invariant Rings

We now want to specialize our discussion to the case that R is a finitely generated Λ algebra.

Problem   Are $T_n(R)$, respectively $\mathscr{J}$, finitely generated?

We will answer this question positively when Λ is a field of characteristic zero in Chapter VI, §5. For the moment we prove a general theorem on the good part of $\mathrm{Spec}(V_n(R))$ over which we have the quotient.

Theorem 4.1   If R is a finitely generated Λ algebra then T_α is finitely generated for all α.

Proof   Let $R = \Lambda[a_1, \ldots, a_k]$. Consider the Azumaya algebra $W_\alpha = \lambda(R)T_\alpha$, with $\alpha = (r_1, \ldots, r_{n^2})$, and let $d = \det \|\mathrm{Tr}(\bar{r}_i \bar{r}_j)\|$. We indicate for simplicity $\bar{r} = \lambda(r)$ when $r \in R$. We know, by construction, that d is invertible in T_α. Let H be the Λ subalgebra of T_α generated by the elements d^{-1}, $\mathrm{Tr}(\bar{r}_i)$, $\mathrm{Tr}(\bar{r}_i\bar{r}_j)$, $\mathrm{Tr}(\bar{r}_i\bar{r}_j\bar{r}_k)$, and $\mathrm{Tr}(\bar{a}_i\bar{r}_j)$. We claim that $H = T_\alpha$. For this purpose let $M = \sum_{i=1}^{n^2} H\bar{r}_i$; M is an H submodule of W_α, free over the elements $\bar{r}_1, \ldots, \bar{r}_{n^2}$. We claim that M is an H subalgebra. In fact, compute $\bar{r}_i\bar{r}_j = \sum \alpha_{ijk}\bar{r}_k$ (for suitable $\alpha_{ijk} \in T_\alpha$). The computation of the α_{ijk} is done in the usual way by

$$\mathrm{Tr}(\bar{r}_i\bar{r}_j\bar{r}_t) = \sum \alpha_{ijk}\,\mathrm{Tr}(\bar{r}_k\bar{r}_t), \qquad t = 1, \ldots, n^2.$$

Solving the system, since $d = \det(\mathrm{Tr}(\bar{r}_k\bar{r}_t))$ and $\mathrm{Tr}(\bar{r}_k\bar{r}_t)$, d^{-1}, $\mathrm{Tr}(\bar{r}_i\bar{r}_j\bar{r}_t) \in M$, we see that $\alpha_{ijk} \in H$. Therefore, M is closed under the product. Similarly we see that $\bar{a}_i \in M$, with $i = 1, \ldots, k$, since $\bar{a}_i = \sum \beta_{ij}\bar{r}_j$ and the β_{ij} are computed solving the system

$$\mathrm{Tr}(\bar{a}_i\bar{r}_t) = \sum \beta_{ij}\mathrm{Tr}(\bar{r}_j\bar{r}_t).$$

Finally it is clear that M is closed under computing the reduced trace, i.e., $\mathrm{Tr}(M) \subseteq M$. It follows that M is an H Azumaya algebra and $M \supset \Lambda[a_1, \ldots, a_k] = \lambda(R)$.

Now $W_\alpha = M \otimes_H T_\alpha$ and for every $r \in R$, $\lambda(r) \in M$. From this, one deduces that the coefficients of the characteristic polynomial of $\lambda(r)$ are in H. Now T_α is generated by these elements and therefore $T_\alpha = H$, in particular T_α is a finitely generated Λ algebra, and we have also estimated the number of elements necessary to generate T_α in terms of k and n.

§5   The Generic Splitting

We apply the preceding discussion to the case that R is itself a rank n^2 Azumaya algebra over Λ. In this case, to give a map of R into a rank n^2 Azumaya algebra B over a commutative ring A is the same as to give an isomorphism $R \otimes_\Lambda A \simeq B$. The set $Q_n(R, A)$ is therefore reduced to one element; in fact it is clear that in this case $U = \mathrm{Spec}(\Lambda)$.

Let us consider instead the functor $\mathscr{M}_\Lambda(R, (A)_n)$. To give an element $\varphi \in \mathscr{M}_\Lambda(R, (A)_n)$ is the same as to give an isomorphism $R \otimes_\Lambda A \simeq (A)_n$.

Proposition 5.1   Based on the preceding statement, $\mathrm{Spec}(V_n(R))$ has a point in A if and only if $R \otimes_\Lambda A$ is isomorphic to $(A)_n$.

Furthermore it is clear that $\mathscr{A}_n(A)$ acts transitively on $\mathscr{M}_\Lambda(R, (A)_n)$ and we see that $\mathrm{Spec}(V_n(R))$ is a homogeneous space over the group $\mathscr{A}_n$. In particular if $\Lambda \to \Gamma$ is any map of commutative rings such that $R \otimes_\Lambda \Gamma \simeq (\Gamma)_n$, we have $V_n(R) \otimes_\Lambda \Gamma \simeq \Gamma(n)$ (in a noncanonical way).

It would be interesting to show that $V_n(R)$ is faithfully flat over Λ. For this purpose it would be sufficient, from the preceding remarks, to show that $\Lambda(n)$ is faithfully flat over Λ. If one can prove this assertion one would have that the generic splitting $R \otimes_\Lambda V_n(R)$ is given by a faithfully flat ring.

§6   Some Special Cases

We have remarked that U_R is a functor in R if we consider only central extension maps.

In particular, if $R \to S$ is surjective we see that U_S is a subscheme of U_R. It is therefore interesting to study the case $R = \Lambda\{x_i\}$, the free ring, since all algebras are quotients of free algebras.

We have already studied in some detail (Chapter III) the generic map $\lambda: R \to (V_n(R))_n$ in this case, and we know that

$$V_n(R) = \Lambda[\xi_{i,st}], \qquad s, t = 1, \ldots, n.$$

Then $\lambda(x_i) = (\xi_{i,st}) = \xi_i$ (a generic matrix).

Let us assume for simplicity that Λ is an infinite field; we can apply to R the conclusion of Theorem **3.2**. In particular we can pass to the quotient field for the following.

Corollary 6.1   Using the terminology of Chapter III, we see that $GL(n, \Lambda)$ acts on the field of fractions of $\Lambda[\xi_{i,st}]$ and the invariant subfield is the center Z of $\Lambda\langle\xi_i\rangle$.

We would like to study more closely the structure of the field Z. For this purpose let $\xi_1 \in \Lambda\langle\xi_i\rangle$ be one of the generic matrices.

Lemma 6.2   The characteristic roots of ξ_1 are algebraically independent over the field Λ.

Proof   It is sufficient to prove that the coefficients of the characteristic polynomial of ξ_1 are algebraically independent. If $\xi_1 = (\xi_{1,st})$ we can specialize the elements $\xi_{1,st}$ as $\xi_{1,st} \to 0$ if $s \neq t$, and for $s = t$, $\xi_{1,ss} \to \xi_{1,ss}$. The matrix ξ_1 specializes to

$$\bar{\xi}_1 = \mathrm{diag}(\xi_{1,11}, \xi_{1,22}, \ldots, \xi_{1,nn})$$

and the characteristic polynomial of ξ_1 specializes to the characteristic polynomial of $\bar{\xi}_1$; and since for this last matrix it is clear that the coefficients of the characteristic polynomial are algebraically independent over Λ, the same is true for ξ_1.

Theorem 6.3   Let $u_1, \ldots, u_n$ be the eigenvalues of ξ_1 and consider the field $F = Z(u_1, \ldots, u_n)$. Then:

   (1) F is a splitting field for the division ring $\Lambda\langle\xi_i\rangle$;
   (2) F is a Galois extension of Z with Galois group $\mathscr{S}_n$, the symmetric group on n elements;
   (3) F is a pure transcendental extension of Λ of transcendence degree $mn^2 - (n^2 - 1)$ (where m = number of free variables x_i).

Proof   (1) Let Ω be an algebraic closure of the field K of fractions of $\Lambda[\xi_{i,st}]$. The map

$$j: R \to (\Lambda[\xi_{i,st}])_n \to (\Omega)_n$$

is an irreducible representation and, since in $(\Omega)_n$ the matrix $j(\xi_1)$ can be brought through an inner automorphism in the diagonal form $\mathrm{diag}(u_1, \ldots, u_n)$, it is clear that this map j is equivalent to a map φ with $\varphi(\xi_1) = \mathrm{diag}(u_1, \ldots, u_n)$.

The map $\varphi: R \to (\Omega)_n$ extends to a map

$$\bar{\varphi}: \Lambda\langle\xi_i\rangle \otimes_Z F \to (\Omega)_n$$

(by Chapter II, **6.8**). Now $\bar{\varphi}(\xi_1) = \mathrm{diag}(u_1, \ldots, u_n)$; therefore, for all $j = 1, \ldots, n$, we have

$$e_{jj} = \prod_{i \neq j}(\bar{\varphi}(\xi_1) - u_i)\prod_{i \neq j}(u_j - u_i)^{-1} \in \bar{\varphi}(\Lambda\langle\xi_i\rangle \otimes_Z F).$$

Now $\bar{\varphi}(\Lambda\langle\xi_i\rangle \otimes_Z F)$ is a simple algebra of degree n containing the n commuting orthogonal idempotents $e_{11}, e_{22}, \ldots, e_{nn}$ and therefore it must split, i.e., $\Lambda\langle\xi_i\rangle \otimes_Z F \simeq (F)_n$. This proves (1).

(2) Consider now the set $\mathscr{L}$ of elements

$$g \in \mathscr{A}_n(\Omega)\left(= \frac{GL(n, \Omega)}{\Omega^*}\right)$$

such that gj maps R in $(F)_n$ and $gj(\xi_1)$ is diagonal. Let σ be a permutation on the elements $1, \ldots, n$, and indicate by A_σ the matrix in $(\Omega)_n$ of this permutation. If $g \in \mathscr{L}$, $A_\sigma g \in \mathscr{L}$, and $g(\xi_1) = \mathrm{diag}(u_1, \ldots, u_n)$, then $A_\sigma g(\xi_1) = \mathrm{diag}(u_{\sigma(1)}, \ldots, u_{\sigma(n)})$. Now if $g \in \mathscr{L}$ then g induces a map $\bar{g} : \Lambda[\xi_{i,st}] \to F$ associated to the morphism j_g:

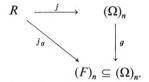

It is clear, by the definition of F, that $\bar{g}(\Lambda[\xi_{i,st}])$ generates F as a field; furthermore, if we modify g by A_σ we have the two maps

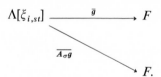

Now A_σ induces an automorphism also of $(\Lambda[\xi_{i,st}])_n$ and the induced map

$$R \to (\Lambda[\xi_{i,st}])_n \xrightarrow{A_\sigma} (\Lambda[\xi_{i,st}])_n$$

is classified by the map

$$\sigma_* : \Lambda[\xi_{i,st}] \to \Lambda[\xi_{i,st}], \qquad \text{where} \quad \sigma_*(\xi_{i,st}) = \xi_{i,\sigma(s)\sigma(t)},$$

i.e., the diagram

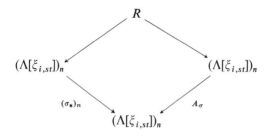

is commutative. Therefore σ_* induces a map $\bar{\sigma}$ to F making the diagram

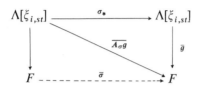

commutative. We have therefore an action of $\mathscr{S}_n$ on F and it acts faithfully on the elements $u_1, \ldots, u_n \in F$ so $\mathscr{S}_n$ is a group of automorphisms of F. Furthermore the field Z, the center of $\Lambda\langle\xi_i\rangle$, is the field of invariants and is always mapped in the same way in F. So $Z \subseteq F^{\mathscr{S}_n}$, and since $F = Z(u_1, \ldots, u_n)$ and the u_i's are the roots of a polynomial of degree n, we have $[F : Z] \leqslant n!$; so $Z = F^{\mathscr{S}_n}$ and $[F : Z] = n!$. This proves (2).

(3) It is clear that we can conjugate the matrix ξ_1 by a matrix B with entries in the algebraic closure K of the field $\Lambda(\xi_{1,st})$ in order to make it diagonal. After conjugation we have $\xi_1{}' = B\xi_1 B^{-1} = \mathrm{diag}(u_1, \ldots, u_n)$. The other matrices become $\xi_2{}', \xi_3{}', \ldots, \xi_m{}', \xi_i{}' = B\xi_i B^{-1}$, and it is clear that if $\xi_i{}' = (\xi_{i,st}')$, the elements $\xi_{i,st}$ obtained by an invertible linear transformation with coefficients in K from the $\xi_{i,st}$ are again algebraically independent over K.

Letting C be the matrix $\mathrm{diag}(\xi_{2,11}', \xi_{2,12}', \ldots, \xi_{2,1n}')$, we see that

$$C\xi_1{}'C^{-1} = \xi_1{}', \ C\xi_i{}'C^{-1} = \xi_i{}''$$

and $\xi_i'' = (\xi_{i,st}'')$ with

$$\xi_{i,st}'' = \xi_{2,1s}'\xi_{i,st}'\xi_{2,1t}';$$

in particular $\xi_{2,1t}'' = \xi_{2,11}'$. We have, therefore, that the ring $\Lambda[\xi_1'', \ldots, \xi_m'']$

is contained in the matrix algebra over the field $\Lambda(u_1, \ldots, u_n, \xi''_{i,st})$ and the $\xi'_{i,st}$ are $[(m - 1)n^2 - (n - 1)]$ different elements. Clearly

$$\Lambda(u_1, \ldots, u_n, \xi''_{i,st})(\xi'_{2,12}, \xi'_{2,13}, \ldots, \xi'_{2,1n}) = \Lambda(u_1, \ldots, u_n, \xi'_{i,st});$$

since the last generators are algebraically independent it follows that the $(m - 1)n^2 + 1$ elements $u_1, \ldots, u_n, \xi''_{i,st}$ are algebraically independent.

Now we finally claim that $\Lambda(u_1, \ldots, u_n, \xi''_{i,st}) = F$; this will complete the proof of (3). Let $G = \Lambda(u_1, \ldots, u_n, \xi''_{i,st})$. Consider the G algebra

$$S = \Lambda\{\xi''_1, \ldots, \xi''_m\}G;$$

it is a simple algebra isomorphic to $\Lambda\langle \xi_i \rangle \otimes_Z G$. We know that

$$e_{ii} = \prod_{i \neq j} (\xi_1' - u_j) \prod (u_i - u_j)^{-1} \in S.$$

Furthermore $e_{ij} = \xi''_{2,ij} e_{ii} \xi''_2 e_{jj} \in S$; therefore $S = (G)_n$. Since $G \supseteq Z$ and $u_1, \ldots, u_n \in G$, we have $F \subseteq G$. $\Lambda\{\xi_i\}F$ is isomorphic to $(F)_n$, but in fact $e_{ii} \in \Lambda\{\xi''_j\}F$ and

$$e_{11}\xi''_2 e_{ii}\xi''_k e_{kk}\xi''_2 e_{11} = \xi''_{2,1i}\xi''_{k,ih}\xi''_{2,h1} e_{11} \in \Lambda\{\xi''_j\}F,$$

so $\xi''_{2,1i}\xi''_{k,ih}\xi''_{2,h1} \in F$; similarly $e_{11}\xi''_2 e_{ii}\xi''_2 e_{11} \in \Lambda\{\xi''_j\}F$ so that $\xi''_{2,1i}\xi''_{2,ij} \in F$. It follows that, since $\xi'_{2,1i} = \xi''_{2,jh}$ for all i, h, we have

$$\xi''_{k,ih} = (\xi''_{2,1i}\xi''_{k,ih}\xi''_{2,hj})(\xi''_{2,1h}\xi''_{2,h1})^{-1} \in F.$$

This finishes the proof of the theorem.

We now ask the following question. Is the center Z of $\Lambda\langle \xi_i \rangle$ a pure transcendental extension of Λ?

In general we can see only this. Consider

$$\Lambda\langle \xi_1, \xi_2 \rangle \subseteq \Lambda\langle \xi_1, \xi_2, \ldots, \xi_m \rangle;$$

we have that this inclusion is a central extension (Chapter II, **6.9.**). If Z_2, Z_m are the centers of the two fields we have

$$Z_2 \subseteq Z_m \quad \text{and} \quad \Lambda\langle \xi_1, \ldots, \xi_m \rangle = \Lambda\langle \xi_1, \xi_2 \rangle Z_m.$$

Proposition 6.4   We claim that if $u_1, \ldots, u_{n^2}$ is a basis of $\Lambda\langle \xi_1, \xi_2\rangle$ over Z_2 then the $n^2(m - 2)$ elements $\text{Tr}(\xi_i u_j)$ $(i = 3, \ldots, m)$ are algebraically independent over Z_2 and generate Z_m over Z_2 (i.e., Z_m is a pure transcendental extension of Z_2).

Proof   By the computation of transcendence degrees of Z_m and Z_2 over Λ, it is enough to show that $F = Z_2(\text{Tr}(\xi_i u_j))$ is equal to Z_m.

Consider $F\Lambda\{\xi_1, \xi_2\} \subseteq \Lambda\langle\xi_1, \ldots, \xi_m\rangle$; it is a simple algebra with center F. If we prove that $\xi_1, \ldots, \xi_m \in F\Lambda\{\xi_1, \xi_2\}$ we will have that $F\Lambda\langle\xi_1, \xi_2\rangle = \Lambda\langle\xi_1, \ldots, \xi_m\rangle$ and therefore $Z = F$.

Now $\xi_i = \sum \alpha_{ij}u_j$ for some $\alpha_{ij} \in Z_m$; taking traces we have $\text{Tr}(\xi_i u_k) = \sum \alpha_{ij}\text{Tr}(u_j u_k)$. As $\text{Tr}(\xi_i u_j)$, $\text{Tr}(u_j u_k) \in F$, so $\alpha_{ij} \in F$ and the theorem is proved.

We finish computing explicitly the field Z_2 for $n = 2$.

Theorem 6.5   The center Z of $\Lambda\langle\xi_1, \xi_2\rangle$ (with ξ_1, ξ_2 two generic 2×2 matrices) is generated by the algebraically independent elements $\text{Tr}(\xi_1)$, $\text{Tr}(\xi_2)$, $\det(\xi_1)$, $\det(\xi_2)$, and $\det(\xi_1 + \xi_2)$.

The elements $1, \xi_1, \xi_2$, and $\xi_1\xi_2$ are a basis of $\Lambda\langle\xi_1, \xi_2\rangle$ over Z with the following multiplication table:

$$\xi_1^2 = \text{Tr}(\xi_1)\xi_1 - \det(\xi_1)$$
$$\xi_2^2 = \text{Tr}(\xi_2)\xi_2 - \det(\xi_2)$$
$$\xi_2\xi_1 = -[\det(\xi_1) + \det(\xi_2) + \det(\xi_1 + \xi_2)]$$
$$+ \text{Tr}(\xi_2)\xi_1 + \text{Tr}(\xi_1)\xi_2 - \xi_1\xi_2.$$

Proof   It is enough to prove the formulas relative to the multiplication table because it follows that, setting

$$F = \Lambda(\text{Tr}(\xi_1), \text{Tr}(\xi_2), \det(\xi_1), \det(\xi_2), \det(\xi_1 + \xi_2)),$$

we find that $F\{\xi_1, \xi_2\}$ is a subalgebra of $\Lambda\langle\xi_1, \xi_2\rangle$ and so $F = Z$. The fact that the five elements considered are algebraically independent follows from the fact that tr deg $Z/\Lambda = 5$.

The first two equations are a consequence of the theorem of Hamilton–Cayley. The last follows from the previous ones and

$$(\xi_1 + \xi_2)^2 = \text{Tr}(\xi_1 + \xi_2)(\xi_1 + \xi_2) - \det(\xi_1 + \xi_2).$$

Remark   This proves the rationality of Z_2 for 2×2 matrices. To try to extend it for $n \times n$ matrices is not such a trivial matter. In fact let us try it for 3×3 matrices, $\Lambda\{x, y\}$ (with x, y two generic 3×3 matrices). Let Z be the center of $\Lambda\{x, y\}$. A basis for $\Lambda\{x, y\}$ over Z is given, for instance, by the elements $x^i y^j$, with $i, j = 0, 1, 2$ (Chapter III, **1.2**). The same computation as before shows that Z contains the elements $\alpha_1, \alpha_2, \alpha_3; \beta_1, \beta_2, \beta_3$ coefficients of the characteristic polynomials of x and y, respectively. Then setting $yx = \sum \gamma_{ij} x^i y^j$, we have nine extra elements. It is easily seen that these 15 elements generate Z over Λ, since we have through these elements the relations necessary to commute y with x and to reduce x^n, y^m, respectively, to combinations of x^2, x, 1; y^2, y, 1. On the other hand, tr deg $Z/\Lambda = 2 \cdot 9 - 8 = 10$, so we must have algebraic relations between these 15 elements. But to try to eliminate five of them does not seem to be an easy task.

Chapter V

FINITELY GENERATED ALGEBRAS AND EXTENSIONS

Throughout this chapter all algebras will have a 1 (although this is not always necessary for the theory to be developed).

§1   Hilbert Nullstellensatz

The purpose of this paragraph is to generalize the Hilbert Nullstellensatz to noncommutative *PI* rings.

Definition 1.1   If F is a field and R an F algebra, we say that R is a Hilbert algebra if the following conditions are satisfied.

(1)   Every primitive ideal has finite codimension.
(2)   Every prime ideal P of R is the intersection of all primitive ideals containing P.

If R satisfies only condition (2), we say that R is a Jacobson ring.

One can interpret condition (2) geometrically as a condition on $\mathrm{Spec}(R)$; in fact, let $\Sigma = \{P \in \mathrm{Spec}(R) | P$ is a primitive ideal$\}$. Then (2) reads as follows. If V is closed in $\mathrm{Spec}(R)$, then $V \cap \Sigma$ is dense in V. In the presence of condition (1) we have, furthermore, that P primitive implies P maximal, so $\Sigma = \{$set of closed points of $\mathrm{Spec}(R)\}$. This is also true in general if R is a PI ring (Chapter II, **1.1**).

The main theorem of this paragraph is the following.

Theorem 1.2 (Hilbert Nullstellensatz)   If R is a finitely generated PI algebra over a field F then R is a Hilbert algebra.

Proof   It is enough to show that every prime ideal P of R is an intersection of maximal ideals of finite codimension.

Consider the ring $R/P = \bar{R}$. Since R satisfies a polynomial identity we have an inclusion $\bar{R} = R/P \to (K)_n$ for a field K and a suitable n (Chapter II, **3.2**). Let $R \to (\tau(R))_n$ be the universal map in matrices; $\tau(R)$ is a finitely generated algebra since R is such (Chapter IV, **1.1**).

We have a commutative diagram

$$
\begin{array}{ccc}
R & \longrightarrow & (\tau(R))_n \\
\downarrow & & \downarrow{\scriptstyle \lambda_n} \\
\bar{R} & \longrightarrow & (K)_n
\end{array}
$$

for a suitable $\lambda: \tau(R) \to K$. Set $A = \lambda(\tau(R))$. We have an inclusion $\bar{R} \to (A)_n$ and A is a finitely generated commutative domain. By the commutative Nullstellensatz we have

$$(0) = \{\bigcap m | m \text{ maximal in } A, A/m \text{ finite dimensional over } F\}.$$

Let

$$U_m = \ker(\bar{R} \to (A)_n \to (A/m)_n),$$

where $\bar{R}/U_m \subseteq (A/m)_n$, and therefore R/U_m is a finite-dimensional algebra over F. Therefore $J(\bar{R}/U_m)$ is the intersection of maximal ideals of finite codimension and $J(R/U_m)^n = 0$.

It follows that if $T = \{\bigcap M \,|\, M$ maximal in R and of finite codimension$\}$ then $T^n \subseteq U_m$ for all m. Now $\bigcap U_m = (0)$ and therefore $T^n = 0$. Since R is a prime ring we must have $T = 0$.

Problem   If R is a finitely generated PI algebra over a field F, is the nil radical of R nilpotent?

This can be proved only in the following somewhat special case.

Proposition 1.3   If R is a finitely generated algebra over a commutative noetherian ring Λ, and if R can be embedded in a Λ algebra S which is a finite module over its center A, then the nil radical of R is nilpotent.

Proof   Let $R = \Lambda\{a_1, \ldots, a_t\}$ and $S = \sum_{i=1}^{m} Au_i$. We can collect a finite number of elements $\alpha_{ij} \in A$ with $i = 1, \ldots, t, j = 1, \ldots, m$, and $\beta_{ijk} \in A$ with $i, j, k = 1, \ldots, m$, such that $a_i = \sum_j \alpha_{ij} u_j$, and $u_i u_j = \sum_k \beta_{ijk} u_k$. The ring $B = \Lambda[\alpha_{ij}, \beta_{ijk}]$ is noetherian and $T = \sum_{i=1}^{m} Bu_i$ is a B algebra such that $R \subseteq T$. Now T is a noetherian ring and so every nil subring of T is nilpotent (cf. 4.4).

One is, therefore, led to investigate which rings R can be embedded in algebras S that are finite modules over their center A.

Let R be such a ring. If, furthermore, R is a finitely generated algebra, we have already found a condition:

(1) $J(R)$ is nilpotent.

Next, if S is a finite module over A, we can embed S in $\mathrm{End}(A)$. If S is generated linearly by n elements over A, then $\mathrm{End}(A)$ is a quotient of a subring of $(A)_n$. Therefore we have found a second condition:

(2) R satisfies all identities of $n \times n$ matrices for some n.

The third necessary condition is more complicated. We recall that if R is a ring and $U \subset R$ is a subset, then $r_R(U) = \{x \in R \,|\, Ux = 0\}$ is called the right annihilator of U and it is a right ideal.

If $R \subset S$ are rings and $U \subset R$, then clearly $r_R(U) = r_S(U) \cap R$. Therefore, if S is right noetherian, then R satisfies the ascending chain condition on right annihilators.

In the course of the proof of **1.3** we have seen that if $R \subseteq S$, which is a finite module over its center A, then R is also contained in a right and left noetherian subring of S. Therefore:

(3) R satisfies the ascending-chain condition on right and left annihilator ideals (and therefore also the descending-chain condition).

The following example, from L. Small [76], shows that condition (3) is independent of the others.

Example (L. Small)   Let F be an infinite field, $F[t]$ the ring of polynomials in t, and $F[t, t^{-1}]$ the ring of Laurent polynomials. Consider

$$A = \left\{ \begin{pmatrix} x & y \\ 0 & z \end{pmatrix} | x, y \in F[t, t^{-1}], \quad z \in F[t] \right\}.$$

A is a subring of $(F[t, t^{-1}])_2$. It is easily seen to be a finitely generated algebra over F. Let

$$I = \left\{ \begin{pmatrix} 0 & y \\ 0 & 0 \end{pmatrix} | y \in F[t] \right\}.$$

I is a right ideal of A. Further, for $i \in \mathbf{Z}$, let

$$y_i = \left\{ \begin{pmatrix} 0 & t^i \\ 0 & 0 \end{pmatrix} \in A \quad \text{and} \quad D_i = y_i A = \begin{pmatrix} 0 & t^i f \\ 0 & 0 \end{pmatrix} | f \in F[t] \right\}.$$

The D_i's form a properly descending chain of right ideals.

We consider now a subring of $(A)_2$, precisely the ring

$$B = \left\{ \begin{pmatrix} \alpha_1 & \alpha_2 \\ 0 & \alpha_3 \end{pmatrix} | \alpha_1 \in F, \quad \alpha_2, \alpha_3 \in A \right\}.$$

B is again finitely generated over F and contained in the 4×4 matrices over $F[t, t^{-1}]$.

$$J(B) = \left\{ \begin{pmatrix} 0 & \alpha \\ 0 & \beta \end{pmatrix} | \alpha \in A, \quad \beta \in J(A) \right\} \quad \text{and} \quad J(B)^3 = 0.$$

Now let

$$\bar{I} = \left\{ \begin{pmatrix} 0 & i \\ 0 & 0 \end{pmatrix} \middle| i \in I \subset A \right\}.$$

$\bar{I}$ is a two-sided ideal of B and so we can form the ring $E = B/\bar{I}$.

E is a finitely generated algebra satisfying all the identities of 4×4 matrices. Furthermore, $\bar{I}^2 = 0$, so $J(E) = J(B/\bar{I}) = J(B)/\bar{I}$, and so $J(E)^3 = 0$.

We now claim that E contains an infinite descending chain of right annihilator ideals. In fact, compute

$$r\left(\begin{pmatrix} 0 & y_i \\ 0 & 0 \end{pmatrix} \right).$$

Now

$$a = \begin{pmatrix} \alpha & \beta \\ 0 & \delta \end{pmatrix} \in B$$

is such that

$$\begin{pmatrix} 0 & y_i \\ 0 & 0 \end{pmatrix} a \in \bar{I}$$

if and only if $y_i \delta \in \bar{I}$.

We verify easily that $y_i \delta \in \bar{I}$ if and only if $\delta \in D_{-i}$. This shows that, passing to E, we find an infinite chain of right annihilators. Therefore, E cannot be embedded in any algebra finite over its center.

It remains an open question whether the three necessary conditions given are also sufficient. The author believes that the answer to this is again no.

§2   Chain Conditions

Let R be a finitely generated algebra over a commutative noetherian ring A. Let n be a natural number and $\mathcal{I}_n = \{I | I$ is an ideal of R and R/I can be embedded in the ring of $n \times n$ matrices over a commutative ring $A\}$; then we have the following.

Theorem 2.1 $\mathscr{I}_n$ satisfies the ascending-chain condition.

Proof   Let $I_1 \subset I_2 \subset I_3 \subset \cdots$ be an ascending chain of ideals in $\mathscr{I}_n$. Let $\lambda\colon R \to (B)_n$, $\lambda_k\colon R/I_k \to (B_k)_n$ be the universal maps in $n \times n$ matrices. By hypothesis λ_k is injective. By universality there are maps $B \to B_1 \to B_2 \to B_3 \to \cdots$ such that the diagram

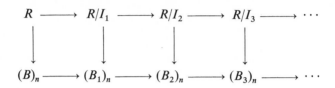

is commutative. Let $J_n = \ker(B \to B_n)$, then $J_1 \subset J_2 \subset J_3 \subset \cdots$ is an ascending chain of ideals in B. Now B is a finitely generated A algebra (Chapter IV, **1.1**) and so it is noetherian. Therefore, the chain stops: $J_k = J_{k+1} = J_{k+2} = \cdots$. Now clearly $I_s = \lambda^{-1}((J_s)_n)$, so $I_k = I_{k+1} = I_{k+2} = \cdots$.

Corollary 2.2   If R is a finitely generated PI algebra over a noetherian ring A, then R satisfies the ascending-chain condition on semiprime ideals.

Proof   Assume that R satisfies an identity of degree d. If I is a semiprime ideal of R we have an injection $R/I \to \oplus_{i=1}^{[d/2]}(B_i)_i$ (Chapter II, **3.1**). Therefore we can embed R/I in $n \times n$ matrices over a commutative ring C, where n is a common multiple between the i's such that $B_i \neq 0$, since if $n = im$, then $((B_i)_i)_m = (B_i)_n$. In particular choose $n = [d/2]!$; we have $I \in \mathscr{I}_n$, and this proves the corollary.

Remark   If R satisfies the condition of the corollary, then R need not satisfy the ascending-chain condition on all ideals even if A is a field.

Example   Take

$$S = \left\{ \begin{pmatrix} a & b \\ 0 & d \end{pmatrix} \middle| a \in F, \quad b, d \in F[x] \right\},$$

with F a field and $F[x]$ the polynomial ring. Take

$$R = \left\{ \begin{pmatrix} u & 0 \\ v & w \end{pmatrix} \middle| u \in F, \quad v, w \in S \right\} \qquad (R \subseteq (F[x])_4).$$

It is easily seen that R is a finitely generated algebra over F. If $W \subseteq F[x]$ is a subspace we set

$$I_W = \left\{ \begin{pmatrix} 0 & 0 \\ m & 0 \end{pmatrix} \middle| m = \begin{pmatrix} 0 & w \\ 0 & 0 \end{pmatrix}, \quad w \in W \right\}.$$

One verifies that I_W is a two-sided ideal. Since $F[x]$ is an infinite-dimensional vector space the ideals I_W do not satisfy the ascending-chain condition.

What is certainly true is the following.

Remark If R is a finitely generated algebra over a commutative noetherian ring Λ and R is a finite module over its center A, then A is finitely generated over Λ (hence A and R are noetherian).

Proof We argue as in **1.3**,

$$R = \sum_{i=1}^{m} A u_i \quad \text{and} \quad R = \Lambda \{ a_1, \ldots, a_t \};$$

$$u_i u_j = \sum \beta_{ijk} u_k \quad \text{and} \quad a_i = \sum \alpha_{ij} u_j, \quad \text{with} \quad \beta_{ijk}, \alpha_{ij} \in A.$$

Consider $B = \Lambda[\alpha_{ij}, \beta_{ijk}]$. We have $B \subset A$ and $R = \sum B u_i$. Hence R is a finite module over the noetherian ring B. Now A is a submodule of R, being the center of R, so A is a finite B module and A is finitely generated over Λ.

We study some consequences of what we have proved.

Proposition 2.3 Let R be a ring satisfying the ascending-chain condition on semiprime ideals. Then every semiprime ideal I is a finite intersection of prime ideals.

Proof   We proceed by contradiction. If there are semiprime ideals which do not satisfy the property of the proposition, we can choose among them a maximal one, call it I. Then (1) I is semiprime; (2) I is not a finite intersection of prime ideals; (3) if $J \supset I$, with J semiprime and $J \neq I$, then J is a finite intersection of prime ideals. From (2) it follows that since I is not a prime ideal there exist ideals J_1, $J_2 \supsetneq I$ such that $J_1 J_2 \subset I$. Now $(J_1 \cap J_2)^2 \subset J_1 J_2 \subset I$ and, therefore, I being semiprime, we have $J_1 \cap J_2 = I$.

Let N_1 and N_2 be the lower nil radicals of J_1 and J_2. N_1 and N_2 are by (3) finite intersections of prime ideals; therefore, $N_1 \cap N_2$ is a finite intersection of prime ideals. We claim that $I = N_1 \cap N_2$. In fact let $r \in N_1 \cap N_2$. Then r is strongly nilpotent modulo J_1 and modulo J_2; i.e., given any sequence $r = r_1, r_2, r_3, \ldots, r_n, \ldots$ with $r_n = r_{n-1} c_{n-1} r_{n-1}$, we must have definitely $r_k \in J_1$, $r_k \in J_2$ for k large. This means that $r_k \in I$ for k large and so it is strongly nilpotent modulo I. Since I is semiprime it follows that $r \in I$. Thus $I = N_1 \cap N_2$ and we have reached a contradiction.

Corollary 2.4   If R satisfies the condition of **2.3** and $I \subseteq R$ is an ideal, then the minimal primes over I are finite in number. If $P_1, \ldots, P_s$ are these minimal primes, then $P_1 \cap P_2 \cap \ldots \cap P_s = N$ is the lower nil radical of I and the decomposition is minimal.

Proof   The proof is like that in the commutative case.

We are now ready to prove an important property of finitely generated PI algebras over noetherian rings.

Theorem 2.5   If R is a finitely generated semiprime PI algebra over a noetherian ring A then R has a total ring of (left and right) quotients, which is a semisimple algebra with descending-chain condition satisfying the same identities as R.

Proof   (0) is a semiprime ideal of R, therefore, from **2.2–2.4** we have $(0) = P_1 \cap P_2 \cap \ldots \cap P_s$ where the P_i's are the minimal prime ideals of R.

Let Q_i be the total ring of quotients of $R_i = R/P_i$, and set $Q = \bigoplus_{i=1}^{s} Q_i$.

Q is clearly a semisimple algebra with descending-chain condition satisfying all identities of R (Chapter II, **5.7**). We have an injection map $\lambda: R \to Q$ defined in the obvious way:

$$\lambda: R = R/\bigcap_{i=1}^{s} P_i \to \bigoplus_{i=1}^{s} R_i \to \bigoplus_{i=1}^{s} Q_i = Q.$$

Identifying R with $\lambda(R)$, we must prove that Q is the (left and right) quotient ring of R. We do it on the left. As the P_i's are all different and minimal, we have $U_i = \bigcap_{j \neq i} P_j \not\subseteq P_i$.

Call V_i the image of U_i in R_i. V_i is a nonzero two-sided ideal of R_i. Furthermore, $U_i \cap P_i = (0)$ so the map $U_i \to V_i$ is bijective. Consider the ring $S = \bigoplus V_i \subseteq R$ (here we go against our convention that all rings should have a 1). We have $S \subseteq R \subseteq Q$. Now if $q_i \in Q_i$ we have $q_i = a_i^{-1} b_i$, with $a_i, b_i \in R_i$ and a_i regular. If $c_i \in V_i$ is a regular element of R_i (Chapter II, **5.5**), we have $(c_i a_i)^{-1}(c_i b_i) = q_i$, with $c_i a_i, c_i b_i \in V_i$. Therefore if $q = (q_1, \ldots, q_s) \in Q$, with $q_i = a_i^{-1} b_i$, we have $q = a^{-1} b$, with $a = (c_1 a_1, c_2 a_2, \ldots, c_s a_s)$, $b = (c_1 b_1, c_2 b_2, \ldots, c_s b_s)$, and $a, b \in S \subseteq R$.

Our final result of finiteness properties is about extensions. Let $R \subset S = R\{a_1, \ldots, a_t\}$ be a finitely generated extension of PI rings (Chapter II, **6.3**). Assume S is prime. We know that R is prime and, if $Q(R)$, $Q(S)$ denote the rings of quotients of R and S, with $Z(R)$, $Z(S)$ the centers of $Q(R)$, $Q(S)$ respectively, we have $Q(R) \subset Q(S)$, $Z(R) \subset Z(S)$. Furthermore if $T = Q(S)^{Q(R)}$ we have $Z(S) \subset T$ where T is a central simple algebra over $Z(S)$ and $Q(S) \simeq Q(R) \otimes_{Z(R)} T$ (Chapter II, **6.8**).

Proposition 2.6 $Z(S)$ is finitely generated as a field over $Z(R)$.

Proof   Let $m_1, \ldots, m_n$ be a basis of T over $Z(S)$. Let $a_i = \sum \alpha_{ij} m_j$, $m_i m_j = \sum \gamma_{ijk} m_k$, with $\alpha_{ij}, \gamma_{ijk} \in Z(S)$.
Consider $F = Z(R)(\alpha_{ij}, \gamma_{ijk})$; we claim that $F = Z(S)$.
Consider $U = \sum F m_i$ where U is an F algebra since $\gamma_{ijk} \in F$. Furthermore $U \otimes_F Z(S) = T$ and so U is a simple algebra with center F and

$$Q(S) = Q(R) \otimes_{Z(R)} T = (Q(R) \otimes_{Z(R)} U) \otimes_F Z(S).$$

We claim that $Q(S) = Q(R) \otimes_{Z(R)} U$; this proves that $F = Z(S)$ and so proves the proposition. Now $R \subset Q(R) \otimes U$ and $a_i \in Q(R) \otimes U$ since

$\alpha_{ij} \in F$. Therefore $S \subset Q(R) \otimes U \subset Q(S)$. As $Q(R) \otimes U$ is a simple algebra, every element of S that is invertible in $Q(S)$ is already invertible in $Q(R) \otimes U$; since $Q(S)$ is generated by S and the inverses of regular elements of S, then $Q(R) \otimes U = Q(S)$.

§3   Jacobson Rings and Hilbert Algebras

We want to generalize the results of §1 to a more abstract setting.

Theorem 3.1   Let $R \subset S = R\{a_1, \ldots, a_k\}$ be a finitely generated extension of prime *PI* rings.

(1)   There is a regular element $c \in R$ such that for every maximal ideal $M \subset R$ with $c \notin M$ we have $MS \cap R = M$.

(2)   If R is a Jacobson ring then for every prime ideal P of R with $c \notin P$, we have $PS \cap R = P$.

Proof   We recall that in an extension $R \to S$, if $I \subset R$ is a two-sided ideal, then IS is a two-sided ideal of S (Chapter II, **6.6**).

Let $Q(R) \subset Q(S)$ be the corresponding extension of quotient rings and let $T = Q(R)\{a_1, \ldots, a_k\} \subset Q(S)$. We have $R \subset S \subset T \subset Q(S)$. Further, let N be a maximal ideal of T, and set $U = T/N$. T is a finitely generated algebra over $Z(R)$, the center of $Q(R)$, and therefore, by **1.2**, T/N is finite dimensional over $Z(R)$. Furthermore $Q(R)$ is a simple algebra, so $Q(R) \cap N = 0$ and we can consider $R \subset Q(R) \subset U$. Now let $\bar{S} = R\{\bar{a}_1, \ldots, \bar{a}_k\}$ be the image of S in U. If we can find a $c \in R$ enjoying the property of the proposition relative to $R \subseteq \bar{S}$, the same c will enjoy this property relative to $R \subseteq S$. So we restrict our consideration to $R \subset \bar{S} \subset U$. If $V = U^{Q(R)}$ we have $Q(R) \otimes_{Z(R)} V \simeq U$. V is finite dimensional over $Z(R)$ and it can be embedded in $(Z(R))_m$ for some number m. Therefore, we embed $Q(R)$ and U in $Q(R) \otimes_{Z(R)} (Z(R))_m \simeq (Q(R))_m$; the embedding gives the elements of $Q(R)$ as constant diagonal matrices. If $q \in (Q(R))_m$ then $q = (q_{ij})$ and, therefore, we can find a regular element $u \in R$ for which $uq_{ij} \in R$. Identifying u with the constant diagonal matrix, we have $uq \in (R)_m$. More generally, given a finite set of elements

$$q_1, \ldots, q_t \in (Q(R))_m,$$

there exists a regular element $u \in R$ with $uq_i \in (R)_m$. Therefore let $b \in R$ be a regular element for which we have $b\bar{a}_i \in (R)_m$, with $i = 1, \ldots, k$. Further, let t be a number such that $t \geq d/2$, where d is the degree of a polynomial identity satisfied by R. We claim that $c = b^t$ is the required element. Assume, therefore, that M is a maximal ideal of R and $c \notin M$. Suppose, by contradiction, that $M\bar{S} = \bar{S}$. Then

$$1 = \sum_{n, \lambda_1, \ldots, \lambda_n} m_{\lambda_1 \ldots \lambda_n} \bar{a}_{\lambda_1} \bar{a}_{\lambda_2} \cdots \bar{a}_{\lambda_n} \qquad (\text{with } \lambda_i = 1, \ldots, k),$$

where $m_{\lambda_1 \ldots \lambda_n} \in M$. Let r be larger then the n's involved. Since $\bar{a}_i b = b\bar{a}_i$ (where $b \in R$ and the a_i's commute with R) we have

$$b^r = \sum m'_{\lambda_1 \ldots \lambda_n} (\bar{a}_{\lambda_1} b)(\bar{a}_{\lambda_2} b) \cdots (\bar{a}_{\lambda_n} b)$$

with $m'_{\lambda_1 \ldots \lambda_n} = m_{\lambda_1 \ldots \lambda_n} b^{r-n} \in M$, $\bar{a}_i b \in (R)_m$.

Therefore, $b^r \in M(R)_m = (M)_m$. This implies that $b^r \in R \cap (M)_m = M$, so b is nilpotent modulo M. Since R/M satisfies an identity of degree d, then R/M is a simple algebra of dimension n^2, with $n \leq [d/2]$ over its center. Therefore, if an element $z \in R/M$ is nilpotent we have $z^n = 0$, and $b^n \in M$. Now $c = b^t \in M$ and $t \geq n$; this is a contradiction.

To prove (2), assume R is a Jacobson ring and $c \notin P$, with P a prime ideal. We have $P = \{\bigcap_{M \supset P} M | M \text{ maximal}\}$ since R is a Jacobson ring. Consider $Q_1 = \{\bigcap M | M \supset P, \text{ with } M \text{ maximal and } c \notin M\}$ and $Q_2 = \{\bigcap M | M \supset P, \text{ with } M \text{ maximal and } c \in M\}$. We have $P = Q_1 \cap Q_2$. Furthermore, $c \in Q_2$ so that $P \neq Q_2$; as P is prime we must have $P = Q_1$. Therefore, $PS = Q_1 S \subset \{\bigcap MS | M \supset P, \text{ with } M \text{ maximal and } c \notin M\}$, and so

$$PS \cap R = Q_1 S \cap R \subset \bigcap_{c \notin M} (MS \cap R) = \bigcap_{c \notin M} M = Q_1 = P.$$

Point (2) is fully proved.

Corollary 3.2   Let $R \subset S = R\{a_1, \ldots, a_k\}$ be a finitely generated extension of prime PI rings. There is a regular element $c \in R$ such that the following properties are satisfied.

(1)   If $M \subset R$ is a maximal ideal and $c \notin M$, there exists a maximal ideal $\tilde{M} \subset S$ with $\tilde{M} \cap R = M$.

(2)   If R is a Jacobson ring and P is a prime ideal of R such that $c \notin P$, then there exists a prime ideal $\tilde{P} \subset S$ with $\tilde{P} \cap R = P$.

Proof   We have to prove only (2), since (1) follows immediately from **3.1**. We know that $PS \cap R = P$. Let $G = \{r \in R | r$ is regular modulo $P\}$. We know that G is a multiplicative set, $PS \cap G = \varnothing$, and P is maximal with respect to the property that $P \cap G = \varnothing$ (Chapter II, **5.6**). Let $\tilde{P}$ be an ideal of S such that $\tilde{P} \supseteq PS$, $\tilde{P} \cap G = \varnothing$, and $\tilde{P}$ is maximal with respect to this property. Then $\tilde{P}$ is a prime ideal and $(\tilde{P} \cap R) \supset P, (\tilde{P} \cap R) \cap G = \varnothing$. Therefore, $\tilde{P} \cap R = P$.

We can interpret this proposition geometrically in terms of Spec. Let us assume that R is a Jacobson ring. We have a map

$$i^* : \operatorname{Spec}(S) \to \operatorname{Spec}(R)$$

and **3.2** tells us that $i^*(\operatorname{Spec}(S)) \supseteq \operatorname{Spec}(R) - V(c)$ which is a dense open set.

Theorem 3.3   Let $R \subset S = R\{a_1, \ldots, a_k\}$ be a finitely generated extension of prime *PI* rings.

(1)   If R is semisimple then S is also semisimple.

(2)   If R is semisimple and S is simple then R is simple and S is finite dimensional over the center of R.

Proof   (1) Consider $Q(R) \subset Q(S)$ and $T = Q(R(\{a_1, \ldots, a_k\})$. T is a prime ring (Chapter II, **6.8**) since $S \subset T$, and $Q(S) = Z(S)S$ so $Q(S) = Z(S)T$. T is a finitely generated prime algebra over $Z(R)$ and so it is semisimple (**1.2**). If, by contradiction, $J(S) \neq 0$, we pick $f \in J(S)$, and $f \neq 0$. We can find a maximal ideal $N \subset T$ such that $f \notin N$. Consider the ring $\overline{T} = T/N$. $\overline{T}$ is a simple ring, finite dimensional over $Z(R)$ (**1.2**), and the image $\bar{f}$ of f in $\overline{T}$ is nonzero. Therefore, $\overline{T} \bar{f} \overline{T} = \overline{T}$ and so we can find elements $m_i, n_i \in \overline{T}$ for which $1 = \sum m_i \bar{f} n_i$. If $\overline{S}$ is the image of S in $\overline{T}$, we have

$$\overline{S} = \{\sum r_{\lambda_1 \ldots \lambda_n} \bar{a}_{\lambda_1} \ldots \bar{a}_{\lambda_n} | r_{\lambda_1 \ldots \lambda_n} \in R\}$$

and

$$\bar{T} = \{\sum q_{\lambda_1 \ldots \lambda_n} \bar{a}_{\lambda_1} \cdots \bar{a}_{\lambda_n} | q_{\lambda_1 \ldots \lambda_n} \in Q(R)\}.$$

Therefore, for any set of elements $g_1, \ldots, g_s \in \bar{T}$, we can find two regular elements $u, w \in R$ such that $ug_i, g_i w \in \bar{S}$. In particular, we choose u, w regular in R such that $um_i \in \bar{S}$ and $n_i w \in \bar{S}$. Let $c \in R$ be a regular element as in 3.1; we have

$$cuw = c(\sum um_i \bar{f} n_i w) \in c\bar{S}\bar{f}\bar{S} \subset J(\bar{S}).$$

As R is semisimple, there is a maximal ideal M in R with $cuw \notin M$. We have $M\bar{S} \cap R = M$, so MS is a proper ideal and can be enlarged to a maximal ideal N of $\bar{S}$, with $N \cap R = M$. We have $cuw \notin N$, but on the other hand, $cuw \in J(\bar{S})$; this is a contradiction. Point (1) is fully proved.

(2) We consider $c \in R$ as in 3.1. Since R is semisimple, there is a maximal ideal M with $c \notin M$. We have that MS is a proper ideal of S and, as S is simple, we must have $MS = 0$. This implies that $M = 0$, and so R is simple. Now R, S simple implies $R = Q(R)$, $S = Q(S)$. Therefore, as $Q(S)$ is a finitely generated algebra over $Q(R)$, it is also finitely generated over $Z(R)$; so by 1.2 we have $Q(S)$ finite dimensional over $Z(R)$.

We arrive finally at the main result of this section.

Theorem 3.4   Let $R \subset S$ be a finitely generated extension of PI rings.

(1) If R is a Jacobson ring we have that S is a Jacobson ring.
(2) If M is a maximal ideal of S, then $R \cap M$ is a maximal ideal of R and S/M is finite dimensional over the center of $R/R \cap M$.
(3) If R is a Hilbert algebra over a field F, then S is also a Hilbert algebra over F.

Proof   (1) Let P be a prime ideal of S. We have to prove that S/P is semisimple. Now S/P is a finitely generated prime extension of $R/P \cap R$, which is semisimple by hypothesis; therefore, 3.3 applies and S/P is semisimple.
(2) This statement follows also from 3.3.
(3) The statement is an immediate consequence of (1) and (2).

It is easy to see, by examples, that the hypothesis that the elements

$a_i \in S$ commute with R is essential for all the theorems proved. For instance, we look at **3.4**. Let

$$R = \left\{ \begin{pmatrix} a & b \\ 0 & c \end{pmatrix} \middle| a, \quad c \in Z, \quad b \in Z_{(p)} \quad \begin{pmatrix} Z_{(p)} \text{ the integers localized} \\ \text{at the prime ideal } (p) \end{pmatrix} \right\}.$$

Let $u = \begin{pmatrix} 0 & 0 \\ 1 & 0 \end{pmatrix}$; the subring of $(Z_{(p)})_2 = S$, generated by R and u, is easily seen to be $(Z_{(p)})_2$. Now R is a Jacobson ring and S is not, of course u does not commute with R.

§4   Dimension Theory

Let R be a ring. As usual we can define the Krull dimension of R by means of $\mathrm{Spec}(R)$.

Definition 4.1
(1)   If P is a prime ideal of R we define rk P (the rank of P) as being the sup of all lengths of chains $P = P_0 \supset P_1 \supset \ldots \supset P_n$ of prime ideals.
(2)   We define dim R (Krull dimension) $= \sup$ rk P.

Our next aim is to prove that for finitely generated PI algebras over noetherian rings, the rank of prime ideals is finite.

Lemma 4.2   Let R be a prime PI ring, P a prime ideal of R, Q the ring of quotients of R, Z the center of Q, and $c \in Z$, $c \neq 0$. Then one of the following two possibilities occurs.

$$I_1 = R \cap PR[c] = P \quad \text{or} \quad I_2 = R \cap PR[c^{-1}] = P.$$

Proof $R[c]$ is a central extension of R and therefore I_1, I_2 are ideals of R. Assume $I_1 \neq P$ and $I_2 \neq P$. We can then find elements $a = \sum_{i=h}^{n} p_i c^i$, $b = \sum_{i=k}^{m} q_i c^{-i}$ regular modulo P. Assume that the choice has been made in such a way that the degree, in c and c^{-1} respectively, of the chosen elements is the minimal possible in both cases. Assume for instance that $n \geq m$. Then

$$ab = \sum_{i=h}^{n} p_i c^i b = \sum_{i=h}^{n-1} p_i b c^i + \sum_{i=k}^{m} p_n q_i c^{n-i}.$$

Now $ab \in I_1$ since $n - i \geq 0$ when $i \leq m$; further, the degree of ab as a polynomial in c is $\leq n - 1$. Therefore, by the minimal choice of n we must have ab not regular modulo P. This is impossible since both a and b are regular modulo P.

Theorem 4.3   Let R, Q, Z, and P be as in the previous lemma. Then there exists a valuation ring A in Z and a prime ideal Q in RA such that $Q \cap R = P$.

Proof   Let A be a subring of Z maximal with respect to the property $PA \cap R = P$. Such an A exists by Zorn's lemma.

Let $S = \{a \in R | a \text{ is regular modulo } P\}$; P is maximal with respect to $P \cap S = \varnothing$, and S is a multiplicative set. Since $PA \cap R = P$ we have $S \cap PA = \varnothing$. If $M \supset PA$ is maximal with respect to $M \cap S = \varnothing$, we have that M is a prime ideal and, further, that $M \cap R = P$, by maximality of P in R with respect to $P \cap S = \varnothing$. Now let $c \in Z$; assume $c \notin A$. If we had $M[c] \cap RA = M$ we would have $PA[c] \cap R \subset M[c] \cap R = M \cap R = P$; this would contradict the maximality of A. Therefore, by **4.2** we must have $PA[c^{-1}] \cap R = P$. Since A is maximal with respect to $PA \cap R = P$, we must have $A[c^{-1}] = A$, i.e., $c^{-1} \in A$. This proves that A is a valuation ring.

Theorem 4.4   Let R be a prime *PI* algebra over a field F. $P \neq 0$ a prime ideal of R. Let Q, $\bar{Q}$ be the rings of quotients of R and $\bar{R} = R/P$. Let Z, $\bar{Z}$ be the centers of Q and $\bar{Q}$, respectively. We have

$$\text{tr deg } \bar{Z}/F \leq \text{tr deg } Z/F.$$

Moreover, if $\text{tr deg } Z/F < \infty$ then $\text{tr deg } \bar{Z}/F < \text{tr deg } Z/F$.

Proof   Let A be a valuation ring in Z, and M a prime ideal of RA such that $M \cap R = P$. Consider $\bar{R} = R/P \subset RA/M = S$; this is a central extension. Let T be the ring of quotients of S, with U its center. We have

$$\begin{array}{c} \bar{R} \subset S \\ \cap \quad \cap, \qquad T = \bar{Q}U, \quad \bar{Z} \subset U. \\ \bar{Q} \subset T \end{array}$$

We claim first that U is algebraic over $\bar{Z}$. Since T is the ring of quotients of S it is enough to show that S is algebraic over $\bar{A}$ (Chapter II, **5.9**). Let $u \in RA$, since $u \in Q(R)$, u satisfies a polynomial $\sum \alpha_i u^i = 0$, with $\alpha_i \in Z$. Since A is a valuation ring in Z, we can multiply all the α_i's by a common factor to make them all in A and one of them invertible in A. Assuming that this is done, we have in $S = RA/M$ that $\sum \bar{\alpha}_i \bar{u}^i = 0$ is a nonzero equation for $\bar{u}$ and so S is algebraic over $\bar{A}$.

At this point we know that

$$\text{tr deg } \bar{Z}/F \leqslant \text{tr deg } U/F = \text{tr deg } \bar{A}/F.$$

Since A is a subring of Z, we have tr deg $\bar{A}/F \leqslant$ tr deg Z/F. Furthermore, if tr deg $Z/F < \infty$ and $A \cap M \neq 0$, we have tr deg $A/A \cap M <$ tr deg Z. We need only prove that $A \cap M \neq 0$. If $u \in M$ is a regular element in R, it satisfies an equation $\sum_{i=0}^{u} \alpha_i u^i = 0$ with $\alpha_i \in A$ and $\alpha_0 \neq 0$; therefore $\alpha_0 = -\sum_{i=1}^{k} \alpha_i u^i \in A \cap M$.

We come now to the main application, i.e., the study of the Krull dimension.

Corollary 4.5   If R is a finitely generated PI algebra over a field F and $P_1, \ldots, P_s$ are the minimal primes, we have dim $R = \max \dim R/P_i$. If R is prime, Q its quotient ring, and Z the center of Q, we have dim $R \leqslant$ tr deg $Z/F < \infty$.

Proof   It is clear that dim $R = \max \dim R/P_i$. We work in the prime case. We know that Z is finitely generated over F (**2.6**), therefore tr deg $Z/F < \infty$. By **4.4** we see immediately that a chain of prime ideals in R can be, at most, of length equal to tr deg Z/F.

We will later prove that dim $R =$ tr deg Z/F by giving a much more precised escription of $\text{Spec}_n(R)$ (where $n^2 = \dim_Z Q$).

§5   Krull Dimension Zero

In this chapter we fix the following symbols: A will be a commutative noetherian ring; $R = A\{a_1, \ldots, a_n\}$ a finitely generated PI algebra. Our purpose is to give a characterization of zero-dimensional algebras of this type (Theorem **5.4**).

We begin with some preliminary material.

Proposition 5.1

(1) If R is semisimple artinian we have that the center Z of R is a finitely generated A algebra.

(2) If R is simple and W is the total field of fractions of $\bar{A} = A \cdot 1 \subset Z$ we have that W is a finitely generated A algebra and R is finite dimensional over W.

Proof   (1)   Assume first that R is simple. In this case Z is a field and we can consider a basis $1 = u_1, u_2, \ldots, u_s$ of R over Z. We have

$$a_i = \sum_j \beta_{ij} u_j, \quad \text{and} \quad u_i u_j = \sum_k \gamma_{ijk} u_k,$$

for suitable $\beta_{ij}, \gamma_{ijk} \in Z$. We claim that setting $\Lambda = A[\beta_{ij}, \gamma_{ijk}]$ we have $Z = \Lambda$.

In fact, $\Lambda \subset Z$ and we can consider the Λ module $S = \sum_{i=1}^{s} \Lambda u_i$. We have $\bar{A} = A \cdot 1 \subset S$, where $a_i \in S$ and S is an A subalgebra since the elements β_{ij} and γ_{ijk} are in Λ. Therefore $S = R$. Since the u_i's form a basis of R over Z and $Z \supset \Lambda$, we must have $\Lambda = Z$. More generally, assume that R is semisimple artinian. We have $R \simeq \oplus_{i=1}^{m} R_i$, with R_i a simple algebra and, if Z_i denotes the center of R_i, with $Z = \oplus Z_i$. Applying our previous result to R_i, which, being a homomorphic image of R, satisfies the hypotheses of the theorem, we see that Z_i is finitely generated over Ae_i, e_i being the unit element of Z_i and R_i. Therefore, Z is generated over A by the elements e_i and the respective generators of the field Z_i over Ae_i.

(2) It is enough to prove that R is finite dimensional over W. In fact if we can prove this we can repeat the same argument as in (1), substituting W for Z and using in the same way the fact that R is finite dimensional over W and finitely generated over A. To see that R is finite dimensional over W, we notice that

$$R = A\{a_1, \ldots, a_k\} = W\{a_1, \ldots, a_k\}.$$

Therefore the claim of (2) follows, since R is simple, from **1.2**.

Proposition 5.2   Let $S = A\{x_1, \ldots, x_h\}$ be a finitely generated algebra over a commutative noetherian ring A. Let M be an ideal of S such that S/M is a right- and left-artinian PI ring. Then M is a finitely generated ideal.

Proof   We separate four cases.

(1) *A is a field.* Let $M_1, \ldots, M_k$ be the maximal ideals of R containing M. Then R/M_i is a simple PI algebra, finitely generated over the field A. By **1.2**, R/M_i is finite dimensional over A. If

$$K_0 = R/M \supset K_1 \supset K_2 \supset \ldots \supset K_s = 0$$

is a composition series of R/M as a right R module, we have that each K_j/K_{j+1} is an irreducible module over one of the rings R/M_i and therefore it is finite dimensional over A. Therefore R/M is finite dimensional over A.

Let $u_1, \ldots, u_s \in S$ be elements such that their classes $\bar{u}_1, \ldots, \bar{u}_s \in S/M$ are a basis modulo M. We must have

$$a_i = x_i - \sum_j \beta_{ij} u_j \in M, \qquad b_{ij} = u_i u_j - \sum_k \gamma_{ijk} u_k \in M$$

for suitable elements $\beta_{ij}, \gamma_{ijk} \in A$. Now let N be the ideal generated by these elements a_i, b_{ij}. We claim that $N = M$.

By construction $N \subset M$. Consider the ring S/N; the subspace W generated over A by the classes in S/N of the elements u_i is a subalgebra since $u_i u_j \equiv \sum_k \gamma_{ijk} u_k \pmod{N}$. Furthermore the classes of the elements x_i belong to W since $x_i \equiv \sum_j \beta_{ij} u_j \pmod{N}$. We must have, therefore,

$W = S/N$ and hence $\dim_A S/N \leqslant \dim_A S/M$. Since $N \subset M$ this implies immediately that $N = M$.

(2) *A is a finite direct sum of fields.* As usual this step is quite simple. Let $A = \oplus_{i=1}^s Z_i$, where Z_i is a field. Let e_i denote the unit element of Z_i. e_i is a central idempotent of S and we have the decompositions

$$S = \oplus Se_i, \quad M = \oplus Me_i, \quad \frac{S}{M} = \oplus \left(\frac{S}{M}\right)e_i \simeq \oplus \frac{(Se_i)}{(Me_i)}.$$

If $Se_i \neq Me_i$, we are under the hypothesis of case (1) and so Me_i is finitely generated over Se_i and also over S. If $Se_i = Me_i$, then Se_i is generated by e_i. Therefore M is a finitely generated ideal.

(3) *S/M is semisimple.* Let S/M be semisimple and Z its center. From **5.1** we know that Z is a finitely generated A algebra. Let $\lambda_1, \ldots, \lambda_k \in S$ be elements which generate, modulo M, Z as an A algebra. Consider the ring $T = A\{\lambda_1, \ldots, \lambda_k\} \subset S$. If $I = ([\lambda_i, \lambda_j])$ is the commutator ideal of T, we have $I \subset M \cap T$ since $\bar{\lambda}_i \in Z$. T/I is a commutative noetherian ring and therefore the ideal $M \cap T/I$ is finitely generated. Let $a_1, \ldots, a_s \in M \cap T$ be generators of this ideal modulo I. Consider in S the ideal N generated by the elements $a_i, [\lambda_i, x_j]$. Clearly $N \subset M$ since $\bar{\lambda}_i \in Z$ and $a_i \in M$. Consider the ring S/N; $S/N \supset \tilde{T} = T/T \cap N$. $\tilde{T}$ is in the center of S/N, since $[\lambda_i, x_j] \in N$ and the x_j's generate S over A; furthermore $T \cap N = T \cap M$ by construction and therefore $\tilde{T} \simeq Z$, so S/N can be considered as a finitely generated Z algebra. Now we can apply to the ideal $M/N \subset S/N$ the conclusion of case (2). We find that M/N is a finitely generated ideal. Since N is a finitely generated ideal it follows that M is also such.

(4) *The general case.* Let $\bar{S} = S/M$ and $J \subset S$ be its Jacobson radical. Assume $J^{n-1} \neq 0, J^n = 0$. Let N be the ideal of S such that $N/M = J^{n-1}$. By case (3) and by induction on the number n we may assume that N is a finitely generated ideal. J^{n-1} is finitely generated both as a right and left R/J module, hence also as a right and left Z module. Therefore, let $c_1, \ldots, c_k \in N$ be generators of N as an ideal. Choose them in such a way that they give generators of J^{n-1} as right and left Z modules. Further, let T be the ideal of S corresponding to J. We have $S/T \simeq \bar{S}/J$. By case (3), let $d_1, \ldots, d_m$ be generators of T as an ideal. We have $c_i d_j \in M$ since $J^n = 0$.

Let $a_1, \ldots, a_p \in S$ be elements that generate, modulo T, Z as an A

algebra. Let $U = A\{a_1, \ldots, a_p\}$; we have, as in case (3), that the kernel I of the map $\varphi \colon U \to Z$ is a finitely generated ideal of U. Let $m_1, \ldots, m_t$ be its generators. We must have $m_j c_i$, $c_i m_j \in M$; $c_i x_k \equiv \sum_j t_{ikj} c_j$, $x_k c_i \equiv \sum_j c_j s_{ikj} \pmod{M}$ for suitable elements $t_{ikj}, s_{ikj} \in U$. Consider the ideal P of S generated by the elements $m_j c_i$, $c_i m_j$, $c_i x_k - \sum_j t_{ikj} c_j$, $x_k c_i - \sum_j c_j s_{ikj}$. We have $P \subset M$. Consider the ring S/P. If we can prove that S/P is right artinian then we will have that the ideal M/P is finitely generated as a right ideal; therefore, since P is finitely generated as an ideal it will follow that M is a finitely generated ideal. To prove that S/P is right artinian, it will be enough, since S/N is right artinian, to show that N/P has a composition series as S module.

Consider N/P as a two sided U module. If we can show that N/P has a composition series as a right U module, it will follow a fortiori that it has a composition series as a right S module. Now consider the elements $\bar{c}_i \in N/P$. It follows from the relation $\bar{c}_i \bar{x}_k = \sum \bar{t}_{ikj} \bar{c}_j$ that $\sum \bar{c}_i S \subset \sum U \bar{c}_j$; similarly from the relations $\bar{x}_k \bar{c}_i = \sum \bar{c}_j \bar{s}_{ikj}$ we have that $\sum S \bar{c}_i \subseteq \sum \bar{c}_j U$. If $u \in I = \ker \varphi \colon U \to Z$, we have $u = \sum_j t_j m_j v_j$ for suitable $t_j, v_j \in U$, so that $\bar{c}_i \bar{u} = \sum_j \bar{c}_i \bar{t}_j \bar{m}_j \bar{v}_j \subset \sum_j U \bar{c}_i \bar{m}_j \bar{v}_j = 0$ since $c_i m_j \in P$. Therefore the right U structure of N/P factors through I and N/P is a right Z module. Further, we claim that the elements $\bar{c}_i$ are a set of generators for such a Z module. As a matter of fact we have, since the elements c_i are a set of generators of N as an ideal,

$$N/P = \sum S \bar{c}_i S \subset \sum S U \bar{c}_i \subset \sum S \bar{c}_i \subset \sum \bar{c}_i U,$$

and therefore $N/P = \sum \bar{c}_i U = \sum \bar{c}_i Z$. Hence N/P is a finite module over Z and therefore has a composition series. This terminates the proof of the theorem.

Proposition 5.3 Let $R = A\{a_1, \ldots, a_k\}$ be a PI algebra over a commutative noetherian ring A, and M a nilpotent ideal of R. If R/M is right and left artinian, then R is right and left artinian.

Proof We have, by the previous proposition, that M is a finitely generated ideal. Let $f_1, \ldots, f_n$ be generators of M. Let J be the radical of R; we will have $J^t = 0$ for some suitable t. Consider the chain $M = M_0 \supset M_1 \supset M_2 \supset \ldots \supset M_k \supset \ldots$ defined inductively by $M_{k+1} = JM_k + M_k J$. We have $M_s = \sum_{a+b=s} J^a M J^b$ and therefore $M_{2t-1} = 0$. We now prove inductively that R/M_s is right and left artinian. As all induction steps are the same, changing our symbols, we can assume that $JM + MJ = 0$. We have $M = \sum R f_i R$; since $JM = MJ = 0$ we can consider M as a two-sided R/J module, where $M = R/J f_i R/J$. We now have $R/J = \bigoplus_{i=1}^{q} R_i$, where R_i is simple with center Z_i and the basis over Z_i is $u_1^{(i)}, \ldots, u_{k_i}^{(i)}$, so that we have $M = \sum Z_i u_j^{(i)} f_k u_s^{(n)} Z_n$. If we prove for $m \in M$ that $Z_i m Z_n$ is a finite-dimensional left Z_n space and a finite-dimensional right Z_n space, we will be finished because it will follow immediately that M has a composition series both as a left and right R module.

Let e_i be the unit of Z_i and consider $Ae_i \subset Z_i$. Let W_i be the field of fractions of Ae_i in Z_i. By **5.1** we have that Z_i is finite dimensional over W_i, say

$$Z_i = \sum_i W_i s_{ji} = \sum_i s_{ji} W_i$$

so that

$$Z_i m Z_n = \sum_{ik} W_i s_{ji} m s_{kn} W_n.$$

We are therefore reduced to prove that $W_i s_{ji} m s_{kn} W_n$ is a finite-dimensional left W_i space and finite-dimensional right W_n space. We shall do one case.

If $w \in W_i$ we have $w = (ae_i)^{-1}(be_i)$ with $a, b \in A$; the inverse is taken in Z_i. Therefore,

$$
\begin{aligned}
w s_{ji} m s_{kn} &= (ae_i)^{-1}(be_i) s_{ji} m s_{kn} = (ae_i)^{-1} b s_{ji} m s_{kn} \\
&= (ae_i)^{-1} s_{ji} m s_{kn} b = (ae_i)^{-1} s_{ji} m s_{kn} b e_n \\
&= (ae_i)^{-1} s_{ji} m s_{kn} (ae_n)^{-1}(ae_n)(be_n) = (ae_i)^{-1} s_{ji} m s_{kn} a(ae_n)^{-1}(be_n) \\
&= s_{ji} m s_{kn} (ae_n)^{-1}(be_n) \in s_{ji} m s_{kn} W_n,
\end{aligned}
$$

unless $ae_n = 0$, in which case,

$$
\begin{aligned}
0 &= (ae_i)^{-1}(be_i) s_{ji} m s_{kn} ae_n = a(ae_i)^{-1}(be_i) s_{ji} m s_{kn} a \\
&= (be_i) s_{ji} m s_{kn},
\end{aligned}
$$

so that $w s_{ji} m s_{kn} = 0$. In any case we see that $W_i s_{ji} m s_{kn} W_n = s_{ji} m s_{kn} W_n$ or it is 0.

After these preliminaries we come to the main result of this ection.

Theorem 5.4   Let $R = A\{a_1, \ldots, a_k\}$ be a *PI* algebra over a commuta-tive noetherian ring A. Then the following conditions are equivalent:

(1) dim $R = 0$;
(2) R is right artinian;
(3) R is left artinian.

Proof   Clearly (2) $\Rightarrow$ (1) and also (3) $\Rightarrow$ (1). Assume that (1) holds. Let N be the nil radical of R. Then $N = \bigcap_{i=1}^{m} P_i$, the P_i's being the mini-mal primes of R (**2.4**). Since dim $R = 0$ we have that all the prime ideals P_i are maximal and therefore $R/N \simeq \oplus_{i=1}^{m} R/P_i$ is right and left artinian. We know that N is nilpotent modulo $L_1(R)$ (Chapter II, **8.1**). Consider the ring $R/L_1(R)$; in this ring we have the ideal $N/L_1(R)$ which is nilpotent and therefore by **5.3** it follows that also $R/L_1(R)$ is right and left artinian. Hence, by **5.2**, $L_1(R)$ is a finitely generated ideal. Now $L_1(R)$ is a sum of nilpotent ideals and, therefore, being finitely generated, it is itself nilpotent. We can then again apply **5.3** and deduce that R is right and left artinian.
    We finally find the real place of polynomial identities for artinian rings.

Proposition 5.5   If R is a ring and M a nilpotent ideal of R, then R satisfies a polynomial identity if and only if R/M satisfies one.

Corollary 5.6   If R is a (left) artinian ring and J its Jacobson radical then R is a *PI* ring if and only if R/J is one.

The proofs of **5.5**, and **5.6** are immediate.

Chapter VI

FINITENESS THEOREMS

§1   Generalities

We study here some applications of the theory developed to the Burnside and Kurosh problems. We recall the two problems.

Definition 1.1
   (1)   A group G is said to be locally finite if every finitely generated subgroup of G is finite.
   (2)   An algebra R over a ring A is said to be locally finite if every finitely generated subalgebra is a finite A module.

The obvious remarks to be made are the following.

Proposition 1.2
   (1)   A locally finite group is a torsion group.
   (2)   A locally finite algebra is integral over A.

   Proof   One applies the local finiteness respectively to the subgroups and the subalgebras generated by one element.

123

The first problem is to see whether the converse of **1.2** is true. The converse of **1.2**(1) is known as the general Burnside problem [44]. The converse of **1.2**(2), in case A is a field, is known as the Kurosh problem [44]. Both problems are known to have a negative answer (cf. [44]).

Here we want to relate these problems with the theory of *PI* rings and show how they have a positive answer if a suitable hypothesis holds. In particular we will show that if a torsion group G is contained in the multiplicative group of a *PI* ring then it is locally finite (cf. **2.8**.).

An integral algebra, with polynomial identity, over a noetherian ring A is locally finite. The assumption that A be noetherian seems indeed unnecessary, but the problem in general is open. We will show how it relates to quite different problems (Hilbert 14th problem) and deduce, from Hilbert's theory, the validity of the above statement for any A containing a field of characteristic zero.*

§2   Integral *PI* Algebras over Noetherian Rings

We recall the main definitions and theorems that are needed.

Definition 2.1
(1)   A commutative noetherian domain A is said to be Japanese if, given any finite dimensional extension F of its field of fractions, the integral closure of A in F is a finite A module.

(2)   A commutative ring A is said to be universally Japanese if it is noetherian and, for every prime ideal P, if A/P is Japanese.

We will abbreviate UJ for universally Japanese and we recall the main facts about UJ rings.

Theorem 2.2
(1)   If A is a field then it is UJ.
(2)   The ring of integers Z is UJ.

---

* The results of this chapter can be improved considerably.
The reader is advised to consult the Addendum to Chapter VI.

(3) If A is UJ then A/I is UJ for any ideal I.
(4) If A is UJ, then $A[x]$, the polynomial ring in one variable, is UJ.
(5) If A is a complete noetherian local ring then A is UJ.

Proof   For the proof see [41, 63].

We start now on our program.

Lemma 2.3   Let R be an A algebra. If $r \in R$ is not integral over A, there exists a prime ideal P of R such that the class of r modulo P is not integral over A.

Proof   Let $S = \{f(r) | f(x) \in A[x]$ runs over all monic polynomials$\}$. By hypothesis $0 \notin S$ and S is clearly a multiplicative set. Therefore, if P is maximal with respect to the property $P \cap S = \varnothing$, we have that P is a prime ideal. If $\bar{r}$ denotes the class of r modulo P, we have $f(\bar{r}) \neq 0$ for every monic polynomial $f(x) \in A[x]$, since $P \cap S = \varnothing$; therefore $\bar{r}$ is not integral over A.

Lemma 2.4   Let $R = A\{x_1, \ldots, x_s\}$ be a finitely generated algebra over a commutative ring A. If $R \otimes A_P$ is a finite A_P module for all prime ideals P of A then R is a finite A module.

Proof   Given a prime ideal P we can find an integer n_P such that $R \otimes A_P$ is generated over A_P by the monomials of degree $\leqslant n_P$. Therefore we can find an $f \notin P$ such that, if M is any monomial of degree $n_P + 1$, we have $fM \in \sum AM_i$, with M_i running over all monomials of degree $\leqslant n_P$. The f's that we find in this way generate clearly the unit ideal A. Let $f_1, \ldots, f_k$ be such that $\sum g_i f_i = 1$, where $f_i M \in \sum_j AM_j$ for every monomial M of degree $n_i + 1$, the M_j's being of degree $\leqslant n_i$. Let $n = \sup n_i$. If M is a monomial of degree $n + 1$ then M can be written as $Q_i N_i$, where Q_i has degree $n_i + 1$. Therefore $f_i M = f_i Q_i N_i = \sum \lambda_{ij} M_j N_i$, with deg $M_j \leqslant n_i$. So $M = \sum g_i f_i M \in \sum AM_\lambda$, with deg $M_\lambda \leqslant n$. This implies by induction that the monomials M span all of R.

These lemmas will be very useful in making various reductions.

Theorem 2.5   Let R be a *PI* algebra over a noetherian ring A. Assume that R is spanned, as an A module, by integral elements over A. Then R is integral over A.

Proof   Assume first that A is a UJ ring. Let us assume by contradiction that there is an element $r \in R$ which is not integral over A. By **2.3** we can find a prime ideal P such that $\bar{r} \in \bar{R} = R/P$ is not integral. $\bar{R}$ satisfies, again, all the conditions of our theorem and, further, is prime; therefore, if we can prove our theorem in the prime case it will hold for any R.

We assume, by the previous reduction, that R is a prime *PI* algebra. Let Q be the ring of quotients of R, Z its center, and $F \subset Z$ the field of fractions of the ring $A1 = \bar{A}$. Further, let U be the subring of Z generated by all traces of elements of R. We generalize the argument of Chapter II, **5.9**. We know (Chapter II, **5.8**) that Z is the field of fractions of U. If $r \in R$ we have $r = \sum \alpha_i r_i$ with r_i integral over A and $\alpha_i \in A$ by assumption; therefore, $\text{Tr}(r_i)$ is integral over A and so $\text{Tr}(r) = \sum \alpha_i \text{Tr}(r_i)$ also is integral over A. Since Z is the field of fractions of U we see that Z is algebraic over F, the field of fractions of $\bar{A}$. Let $r \in R$ be any element, let $c_1, \ldots, c_k \in R$ be a basis of Q over Z; there is one basis since $Q = RZ$ (Chapter II, **5.7**). Consider the algebra $S = A\{c_1, \ldots, c_k, r\}$ with $r \in R$. We have clearly $SZ = Q$ since the elements c_i are a basis of Q over Z. Therefore $S \subseteq Q$ is a central extension and S is a prime ring (Chapter II, **6.7**).

Let us call T the ring of quotients of S, and W its center. We have $S \subset T \subset Q$ and $\bar{A} \subset W \subset Z$ (Chapter II, **6.8**). Since S is a finitely generated algebra over A, we have that W is finitely generated as a field over F (Chapter V, **2.6**). Now $W \subset Z$ and Z is algebraic over F, therefore W is finite dimensional over F.

Let C be the integral closure of $\bar{A}$ in W; since A is UJ we know that C is a finite A module. We have previously seen that if $u \in R$, then $\text{Tr}(u)$ is integral over $\bar{A}$. Since $Q = T \otimes_W Z$ we must have, for every $u \in S$, that $\text{Tr}(u) \in W$ and, being $\text{Tr}(u)$ integral over $\bar{A}$, that $\text{Tr}(u) \in C$.

Consider now $u \in S$ expressed through the basis $c_1, \ldots, c_k$ over W:

$$u = \sum \alpha_i c_i, \qquad \alpha_i \in W.$$

We can compute the elements α_i with the standard formula:

$$\text{Tr}(uc_j) = \sum_i \alpha_i \text{Tr}(c_i c_j).$$

This is a system of equations with $d = \det(\mathrm{Tr}(c_ic_j)) \neq 0$. We can thus solve this system and we see, since $\mathrm{Tr}(uc_j)$, $\mathrm{Tr}(c_ic_j) \in C$ (as uc_j, $c_ic_j \in S$), thet $\alpha_i \in (1/d)C$. Therefore we have just computed that $S \subset \sum_i (1/d)Cc_i$. Now $\sum(1/d)Cc_i$ is a finite C module, hence a finite A module. Since S is a submodule of this module and A is noetherian, we deduce that S is also a finite A module. Therefore $r \in S$ is integral over A. Now r was an arbitrary element of R and so the theorem is proved if A is UJ.

Let A be noetherian, $r \in R$. Let $S = A[r]$; we have to prove that S is a finite A module. If P is a prime ideal of A we have $S_P \subseteq R_P$ as algebras over A_P, and S is a finite A module if and only if S_P is a finite A_P module for all P. Now consider the completion $\hat{A}_P$ of A_P. $\hat{A}_P$ is a faithfully flat extension of A_P and so S_P is a finitely generated A_P module if and only if $\hat{S}_P = S_P \otimes_{A_P} \hat{A}_P$ is a finitely generated $\hat{A}_P$ module. Since $\hat{R}_P = R_P \otimes_{A_P} \hat{A}_P$ is spanned over $\hat{A}_P$ by integral elements, then by the previous case, and since $\hat{A}_P$ is a UJ ring, we know that $\hat{R}_P$ is made of integral elements. Therefore $\hat{S}_P = \hat{A}_P[r]$ is a finite $\hat{A}_P$ module and the theorem is proved for any noetherian ring ($\hat{S}_P \subseteq \hat{R}_P$ because $\hat{A}_P$ is faithfully flat over A_P). (I owe this final argument to S. Greco.)

We now come to the Kurosh problem.

Lemma 2.6   Let $R = A\{a_1, \ldots, a_k\}$ be a finitely generated algebra over a commutative noetherian ring A. If R is not a finite A module there exists a prime ideal P such that R/P is not a finite A module.

Proof   Let $\mathcal{J} = \{I | I \text{ is an ideal of } R \text{ and } R/I \text{ is not a finite module}\}$; we want to prove that $\mathcal{J}$ satisfies the conditions that allow us to use Zorn's lemma. Let $\{I_\alpha\}_\alpha$ be an increasing family of ideals in $\mathcal{J}$ and let $I = \bigcup I_\alpha$. Assume by contradiction that $I \notin \mathcal{J}$, so that $\bar{R} = R/I$ is a finite A module. Let $r_1, \ldots, r_t \in R$ be such that $\sum_i A\bar{r}_i = \bar{R}$. We must have $\bar{a}_i = \sum_j \alpha_{ij}\bar{r}_j$ and $\bar{r}_i\bar{r}_j = \sum \beta_{ijk}\bar{r}_k$ for suitable $\alpha_{ij}, \beta_{ijk} \in A$. Therefore

$$u_i = a_i - \sum \alpha_{ij}r_j, \; v_{ij} = r_ir_j - \sum_k \beta_{ijk}r_k \in I.$$

Since $I = \bigcup I_\alpha$ there exists an index α_0 such that $u_i, v_{ij} \in I_{\alpha_0}$. Consider the ring $\tilde{R} = R/I_{\alpha_0}$; in $\tilde{R}$ the submodule $U = \sum A\tilde{r}_i$. We have $\tilde{a}_i = \sum \alpha_{ij}\tilde{r}_j \in U$ and further U is a subalgebra since $\tilde{r}_i\tilde{r}_j = \sum \beta_{ijk}\tilde{r}_k$. Therefore, as $\tilde{R}$ is generated as an algebra by the elements $\tilde{a}_i$, we have that $U = \tilde{R}$ and therefore $\tilde{R}$ is a finite module; this contradicts the hypothesis $I_{\alpha_0} \in \mathscr{J}$.

If R itself is not a finite A module, the set $\mathscr{J}$ is not empty and, by what we have just proved, we can find an ideal P maximal in $\mathscr{J}$. R/P is not a finite A module by construction. We wish to prove that P is a prime ideal. Assume therefore, by contradiction, that there exist two ideals I, $J \supset P$, with $I, J \neq P$, such that $IJ \subset P$. Consider $I \cap J = K$; we have $K \supset P$ and $K^2 = (I \cap J)^2 \subset IJ \subset P$.

We split the discussion to consider two cases. First, assume $K = P$, then $R/P \subset R/I \oplus R/J$ and R/I, R/J are finite A modules by the maximality of P. Then $R/P \subset R/I \oplus R/J$ is also a finite A module since A is noetherian. In the second case we assume $K \neq P$. Then by maximality of P, $R/K = \bar{R}$ is a finite A module. Let $b_1, \ldots, b_s \in R$ be such that $\bar{R} = \sum_{i=1}^{s} A\bar{b}_i$. As before we have $\bar{a}_i = \sum_j \alpha_{ij}\bar{b}_j$, $\bar{b}_i\bar{b}_j = \sum_k \beta_{ijk}\bar{b}_k$ for suitable α_{ij}, $\beta_{ijk} \in A$ and so

$$u_i = a_i - \sum_j \alpha_{ij}b_j, v_{ij} = b_ib_j - \sum_k \beta_{ijk}b_k \in K.$$

Let M be the ideal of $\tilde{R} = R/P$ generated by the classes $\tilde{u}_i$, $\tilde{v}_{ij}$. By the same argument as before, $\tilde{R}/M$ is spanned by the classes of the elements $\tilde{b}_i$ and so it is a finite A module. On the other hand,

$$M = \sum \tilde{R}\tilde{u}_i\tilde{R} + \sum \tilde{R}\tilde{v}_{ij}\tilde{R}.$$

Since $M \subset \tilde{K}$ and $\tilde{K}^2 = 0$, we have

$$M = \sum \frac{\tilde{R}}{\tilde{K}\tilde{u}_i}\frac{\tilde{R}}{\tilde{K}} + \sum \frac{\tilde{R}}{\tilde{K}\tilde{v}_{ij}}\frac{\tilde{R}}{\tilde{K}} = \sum A\tilde{b}_j\tilde{u}_i\tilde{b}_k + \sum A\tilde{b}_k\tilde{v}_{ij}\tilde{b}_s.$$

We have thus proved that M is a finite A module. Since R/M is also a finite A module, this is a contradiction and therefore we have proved that P is a prime ideal.

Theorem 2.7   Let $R = A\{a_1, \ldots, a_k\}$ be a finitely generated PI algebra over a noetherian ring A. If R is integral over A then R is a finite A module.

Proof   We assume first that A is a UJ ring. We work by contradiction; using the previous proposition we may assume, passing to R/P for a prime ideal P, that R is a prime ring. As usual, let Q be its quotient ring, Z the center of Q, and F the field of fractions of $\bar{A} = A1 \subset Z$. We argue essentially as in **2.5**. First, as in **2.5**, it follows that Z is algebraic over F and, as R is finitely generated over $\bar{A}$, that Z is finite dimensional over F. Let C be the integral closure of $\bar{A}$ in Z. Since A is UJ we know that C is a finite A module. If $c_1, \ldots, c_t \in R$ form a basis of Q over Z, we have for any $r \in R$, that $r = \sum \alpha_i c_i$, with $\alpha_i \in Z$, and $\mathrm{Tr}(rc_j) = \sum \alpha_i \mathrm{Tr}(c_i c_j)$. Since R is integral over A, it follows that $\mathrm{Tr}(rc_i)$, $\mathrm{Tr}(c_i c_j) \in C$. Therefore, if $d = \det(\mathrm{Tr}(c_i c_j))$, then $d \neq 0$ and we have $R \subset \sum (1/d)Cc_i$. Finally, since A is noetherian and $\sum (1/d)Cc_i$ is a finite A module, it follows that R is a finite A module.

If A is any noetherian ring, one sees that for any prime ideal P of A the ring $R \otimes_A \hat{A}_P$ is a finitely generated algebra over a UJ ring spanned by integral elements. Therefore $R \otimes_A \hat{A}_P$ is a finitely generated $\hat{A}_P$ module for every P. It follows as in **2.5** that R is a finite A module (cf. **2.4**).

Corollary 2.8

(1) (Bounded Kurosh problem) If R is an integral algebra of bounded degree (Chapter I, **3.21**) over a noetherian ring A, then R is locally finite.

(2) (Special case of Burnside problem) If G is a group of invertible elements of R, a *PI* algebra, and every element of G has finite order, then G is locally finite.

Proof (1) Use Chapter I, **3.22** and **2.7**.

(2) Let $g_1, \ldots, g_k \in G$ and consider the subring S of R generated by the g_i's, i.e., $S = Z\{g_1, \ldots, g_k\} \subseteq R$. S is spanned by the monomials in the g_i's and these are elements of G. Each such element satisfies $g^n = 1$ for some n and therefore satisfies an integral equation over Z. Therefore **2.5** and **2.7** apply and S is a finite Z module. Call M the subgroup generated by the elements g_i. If M is not finite consider the set of ideals I of S such that $\bar{M} \subset S/I$ is not finite. Since S is noetherian (being a finite Z module) there is a maximum such I. We claim that I is a prime ideal. First, $\bar{S} = S/I$ is torsion free, since, if T is the torsion of S, then T is finite; therefore the image of M in $\bar{S}/T$ is finite if and only if the image of M in S is finite. We split the discussion to two cases as in **2.6**. If $(0) = J_1 \cap J_2$ in S with $J_1 \neq (0)$, $J_2 \neq 0$, we have an injection $\bar{S} \rightarrow \bar{S}/J_1 \oplus \bar{S}/J_2$ and since the images of M in $\bar{S}/J_1$ and $\bar{S}/J_2$ are finite, then by maximality of I, M must be finite modulo I. If, instead, one has an ideal $J \neq (0)$ in $\bar{S}$ with $J^2 = 0$, we consider the map $\bar{S} \rightarrow \bar{S}/J$ and the images M_1, M_2 of M in $\bar{S}$, $\bar{S}/J$, respectively. The kernel of the map $M_1 \rightarrow M_2$ is made of elements of the form $1 + j$, $j \in \bar{J} = J/I$; now the map $\bar{J} \rightarrow$ Units of $\bar{S}$, given by

$j \to j + 1$, is a group monomorphism since $J^2 = 0$. But J is torsion free and so the kernel of $M_1 \to M_2$ is a torsion-free periodic group, hence it is (0). Since M_2 is finite, M_1 is also finite. Finally, by contradiction we may assume that $\bar{S}$ is a prime ring and torsion free over Z. Now $\bar{S} \otimes_Z Q$ is a simple algebra of degree p^2 over its center F for some p, spanned by the image $\bar{M}$ of M in $S \otimes_Z Q$. Let $g_1, \ldots, g_n \in \bar{M}$ be a basis of $\bar{S} \otimes_Z Q$ over its center F. If $g \in \bar{M}$ we have $g = \sum \alpha_i g_i$, $\alpha_i \in F$; now α_i is computed solving the equations $\mathrm{Tr}(g g_j) = \sum \alpha_i \mathrm{Tr}(g_i g_j)$. The elements $g g_i$, $g_i g_j \in \bar{M}$ are in $\bar{M}$, and for $h \in \bar{M}$ we have $h^m = 1$ for some exponent m. The characteristic roots of h are therefore mth roots of 1. On the other hand the characteristic roots of h satisfy the characteristic polynomial of h, which is of degree n. Now it is known that the roots of 1 which are algebraic of degree $\leqslant n$ are a finite set. Therefore the numbers α_i, which are suitable sums of n elements out of this finite set, themselves form a finite set. This proves that M is a finite set, reaching a contradiction.

The results obtained so far all lack one property that would make the commutative case much clearer. In fact if $A \subset B$ are commutative rings and $B = A[b_1, \ldots, b_s]$ then the assumption that the elements b_i are integral over A implies by itself that B is a finite A module and in fact one can give a priori a set of generators for this module using only the degrees of the integral polynomials satisfied by the elements b_i. Clearly a result of this strength cannot exist in noncommutative rings even if very strongly restricted. For instance if $R = (F[x])_2$, the two-by-two matrices over the polynomial ring $F[x]$ over a field F, and

$$a_1 = \begin{pmatrix} 0 & x \\ 0 & 0 \end{pmatrix}, \qquad a_2 = \begin{pmatrix} 0 & 0 \\ x & 0 \end{pmatrix}$$

then $a_1^2 = 0$, $a_2^2 = 0$, but $F\{a_1, a_2\}$ is not integral over F since

$$a_1 a_2 = \begin{pmatrix} x^2 & 0 \\ 0 & 0 \end{pmatrix}$$

is clearly not integral.

On the other hand one may hope that, even in the noncommutative case, one may pass from an infinite number of conditions (all elements of R are integral over A) to a finite number and still **2.7** would be true. We want to discuss this interesting point; we start with a lemma and a theorem valid for any field A.

Lemma 2.9   Let $\mathbf{Z}\{x_1, \ldots, x_k\}$ be the free algebra in k variables over the integers and m a natural number. There exists a *finite* number of monomials $M_1, \ldots, M_s$ in the variables x_i with the following property: If $\varphi: \mathbf{Z}\{x_1, \ldots, x_k\} \to (K)_n$ is an irreducible representation with K a field and $n \leqslant m$, we have $\sum_i \varphi(M_i)K = (K)_n$.

    Proof   Choose $n \leqslant m$. Let

$$\lambda: \mathbf{Z}\{x_1, \ldots, x_k\} \to (\mathbf{Z}[x_{t,ij}])_n = (A)_n$$

be the universal map in $n \times n$ matrices, $\lambda(x_t) = \xi_t = \sum x_{t,ij}e_{ij}$. For every representation

$$\varphi: \mathbf{Z}\{x_1, \ldots, x_k\} \to (K)_n$$

we have a unique map $\bar{\varphi}: A \to K$ such that the following diagram is commutative:

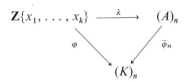

The map $\bar{\varphi}_n$ is such that the following diagram is commutative:

$$
\begin{array}{ccc}
(A)_n & \xrightarrow{\ \mathrm{Tr}\ } & A \\
\Big\downarrow{\scriptstyle\bar{\varphi}_n} & & \Big\downarrow{\scriptstyle\bar{\varphi}} \\
(K)_n & \xrightarrow{\ \mathrm{Tr}\ } & K
\end{array}
$$

    If $\mathbf{a} = (N_1, \ldots, N_n)$ is an n^2-tuple of monomials in x_i, we will have $\sum_i \varphi(N_i)K = (K)_n$ if and only if $\Delta(\mathbf{a}) = \det(\mathrm{Tr}\ \varphi(N_i)\varphi(N_j)) \neq 0$. Consider $d(\mathbf{a}) = \det(\mathrm{Tr}(\lambda(N_i)\lambda(N_j)))$; clearly we have $\bar{\varphi}(d(\mathbf{a})) = \Delta(\mathbf{a})$. Let I be the ideal of A generated by all the elements $d(\mathbf{a})$, where $\mathbf{a}$ varies among all possible n^2-tuples of monomials in the variables x_i. We have just seen that φ is an irreducible representation if and only if $\bar{\varphi}(I) \neq 0$.

    Now A is a noetherian ring and therefore I is finitely generated. Say that $d(\mathbf{a}_1), \ldots, d(\mathbf{a}_s)$ generate I for some s n^2-tuples, $\mathbf{a}_1, \ldots, \mathbf{a}_s$.

From what we have seen $\varphi: \mathbf{Z}\{x_1, \ldots, x_k\} \to (K)_n$ is an irreducible representation if and only if $\bar{\varphi}(d(\mathbf{a}_i)) \neq 0$ for one suitable $\mathbf{a}_i$. Consider then the $s \cdot n^2$ monomials appearing in the n^2-tuples $\mathbf{a}_1, \ldots, \mathbf{a}_s$. We have clearly proved that their images span every irreducible representation $\varphi: \mathbf{Z}\{x_1, \ldots, x_k\} \to (K)_n$.

Repeat the argument for all $n \leqslant m$ and collect all the monomials obtained for each n; we have in this way a finite number of monomials satisfying the conditions of the lemma.

Theorem 2.10   Let $\mathbf{Z}\{x_1, \ldots, x_k\}$ be the free algebra in k variables. Given a natural number m, there exists a finite number of monomials $N_1, \ldots, N_t$ in the x_i's such that the following statement is true.

Given any algebra $R = F\{a_1, \ldots, a_k\}$ over a field F of degree $\leqslant m$ (Chapter II, **7.5**), consider the map $\lambda: \mathbf{Z}\{x_1, \ldots, x_k\} \to R$ given by $\lambda(x_i) = a_i$; then R is finite dimensional over F if and only if the elements $\lambda(N_i)$ are algebraic over F.

Proof   Let $M_1, \ldots, M_s$ be the elements considered in the previous lemma relative exactly to m. We take, for the N_i's of the statement, exactly the monomials $x_j M_i$, $M_i M_j$, $M_i M_j M_k$. We wish to prove that they satisfy the conditions of the theorem. We notice that if $m = 1$, in particular if R is commutative, the M_i's can be taken to be just the monomial 1 so that the N_i's turn out to be the elements x_i and we obtain the familiar statement that a finitely generated commutative algebra is finite if and only if it is generated by algebraic elements.

Now let $R = F\{a_1, \ldots, a_k\}$ be an algebra of degree $\leqslant m$ over a field F and let $\lambda: \mathbf{Z}\{x_1, \ldots, x_k\} \to R$ be as above, with $\lambda(x_i) = a_i$. If R is finite dimensional over F then every element of R is algebraic over F. Assume conversely that the elements $\lambda(N_i)$ are algebraic over F. If by contradiction R were not finite dimensional over F then we could find a prime ideal P such that $\bar{R} = R/P$ is still not finite dimensional over F (**2.6**). So we can assume, using this reduction, that R is prime. Let Q be the ring of quotients of R, and $\mathscr{Z}$ the center of Q, with $\bar{\mathscr{Z}}$ an algebraic closure of $\mathscr{Z}$. We have $Q \otimes \bar{\mathscr{Z}} \simeq (\bar{\mathscr{Z}})_n$ with $n \leqslant m$, since R is of degree $\leqslant m$.

Consider the representation

$$\varphi: \mathbf{Z}\{x_1, \ldots, x_k\} \xrightarrow{\lambda} R \to Q \to Q \otimes \bar{\mathscr{Z}} \simeq (\bar{\mathscr{Z}})_n.$$

Since $R = F\lambda(\mathbf{Z}\{x_1, \ldots, x_k\})$ and $Q = R\mathscr{Z}$ we have that φ is an irreducible representation, therefore we will have $\sum \varphi(M_i)\overline{\mathscr{L}} = (\overline{\mathscr{L}})_n$. Identifying R with the corresponding subring of $(\overline{\mathscr{L}})_n$ we have $\varphi(M_i) = \lambda(M_i) \in R$ and therefore $\sum_i \lambda(M_i)\mathscr{L} = Q$.

Now let $P_1, \ldots, P_{n^2}$ be a basis of Q over $\mathscr{L}$ chosen out of the elements $\lambda(M_i)$. Our hypothesis implies that the elements $a_i P_j$, $P_i P_j$, $P_i P_j P_k$ are algebraic over F. Therefore $\mathrm{Tr}(a_i P_j)$, $\mathrm{Tr}(P_i P_j)$, $\mathrm{Tr}(P_i P_j P_k)$ are algebraic over F. Let $W \subset \mathscr{L}$ be the field generated over F by these elements. Then W is clearly finite dimensional over F. Consider $a_i = \sum_j \alpha_{ij} P_j$ for suitable $\alpha_{ij} \in \mathscr{L}$. We solve for the α_{ij} and obtain the equations

$$\mathrm{Tr}(a_i P_k) = \sum_j \alpha_{ij} \mathrm{Tr}(P_j P_k).$$

Since $\mathrm{Tr}(a_i P_k)$, $\mathrm{Tr}(P_i P_j) \in W$ and $d = \det(\mathrm{Tr}(P_i P_j)) \neq 0$, we have $d \in W$ and so $\alpha_{ij} \in W$. Furthermore $P_i P_j = \sum_k \beta_{ijk} P_k$ for suitable $\beta_{ijk} \in \mathscr{L}$ and again we solve

$$\mathrm{Tr}(P_i P_j P_n) = \sum \beta_{ijk} \mathrm{Tr}(P_k P_n).$$

Since $\mathrm{Tr}(P_i P_j P_n) \in W$ we still have $\beta_{ijk} \in W$.

What we have proved is that $T = \sum_i W P_i$ is a subalgebra of Q and that $a_i \in T$ and therefore $R \subset T$. Now T is finite dimensional over F, as W is, so this implies that R is also finite dimensional over F. This proves the theorem.

This theorem is a strong generalization of **2.7** for a field A, but still does not allow us to make predictions on the size of the dimension of R starting from the degrees of the algebraic equations satisfied by the elements N_i. We will see that it is necessary to drop the assumption that A is a field to get an answer to this, but we will not be able to get an answer on the problem stated.

Before passing to this more difficult step we notice some theorems similar to those proved before, where "integral" or "algebraic" is replaced by "nil."

Theorem 2.11   Let R be a *PI* algebra over a ring A. Assume that R is spanned as an A module by nilpotent elements. Then R is nil.

Proof   Assume that R is not nil. Then there is a prime ideal P of R such that R/P is nonzero and satisfies the hypothesis of the theorem. We may assume by contradiction that R is prime. Let Q be the ring of quotients of R, and Z the center of Q. Since $Q = RZ$ and R is spanned by nilpotent elements, we will have that Q is also spanned over Z by nilpotent elements. Therefore if $q \in Q$ we have $q = \sum \alpha_i q_i$, with $\alpha_i \in Z$ and q_i nilpotent. Computing traces we obtain $\text{Tr}(q) = \sum \alpha_i \text{Tr}(q_i) = 0$. This is a contradiction since $\text{Tr}(ab)$ is a nondegenerate bilinear form on Q.

Corollary 2.12   Given that R is a PI algebra over a ring A with a proper identity of degree d and $S \subseteq R$ is a nil semigroup of R.

(1)   Then $S^{[d/2]} \subseteq L_1(R)$.
(2)   If S is finitely generated it is finite.

Proof   (1)   Consider $AS = \{\sum a_i s_i | a_i \in A, \ s_i \in S\}$. AS is a nil subalgebra by **2.11**; therefore by Chapter II, **8.1** we have $(AS)^{[d/2]} \subseteq L_1(R)$ and so (1) is proved.

(2)   If S is finitely generated then the nil subalgebra that it generates is nilpotent, so for some $N > 0$ we have $S^N = 0$. In particular the monomials in the generators of S of degree $\leqslant N$ yield all the elements of S.

Again in the contest of nilpotency we have the following result which replaces the commutative theorem: If $a_i^n = 0$, $i = 1, \ldots, m$, then the ring generated by a_i's is nilpotent of degree $\leqslant \sum n_i - m + 1$.

Theorem 2.13   Given integers m and k, there exist monomials $N_1', \ldots, N_t'$ of positive degree in $\mathbf{Z}^+\{x_1, \ldots, x_k\}$ such that if $R = \mathbf{Z}^+\{a_1, \ldots, a_k\}$ is a PI ring of degree $\leqslant m$, then R is nilpotent if and only if the monomials $N_1', \ldots, N_t'$ computed in the elements $a_1, \ldots, a_k$ are nilpotent. Furthermore the index of nilpotency of R is bounded by a number q depending only on the identities of R and on the nilpotence degrees of $N_1', \ldots, N_t'$, and not on R.

Proof   We repeat the arguments of **2.9** relative to m, but assume that $\varphi : \mathbf{Z}^+\{x_1, \ldots, x_k\} \to (K)_n$ (with $n \leqslant m$) is irreducible (we drop 1 from the free algebra and consider only the positive part of it). We are thus able to find a finite number of monomials of positive degree $N_1', \ldots, N_t'$ with the same properties as those of **2.9** relative to this subclass of irreducible representations. We claim that they satisfy the requirements of the theorem. For this let $\psi\colon \mathbf{Z}^+\{x_1, \ldots, x_k\} \to \mathbf{Z}^+\{a_1, \ldots, a_k\}$ be the map $x_i \to a_i$ and assume that $\psi(N_i')$ is nilpotent. If $R = \mathbf{Z}\{a_1, \ldots, a_k\}$ were not nil then we could find a proper prime ideal $P \subset R$. Therefore we could construct (Chapter II, **2.4**) the maps

$$\mathbf{Z}^+\{x_1, \ldots, x_k\} \overset{\psi}{\to} R \to R/P \to Q \to Q \otimes_Z \bar{Z} \simeq (\bar{Z})_n,$$

with Q the quotient ring of R/P, Z its center, $\bar{Z}$ the algebraic closure, and $n \leqslant m$ a suitable number.

This yields an irreducible representation $\varphi\colon \mathbf{Z}^+\{x_1, \ldots, x_k\} \to (\bar{Z})_n$ of degree $n \leqslant m$, so $\sum \bar{Z}\varphi(N_i') = (\bar{Z})_n$. This is clearly a contradiction since the $\varphi(N_i')$'s are nilpotent.

To show that there is an a priori estimate of the nilpotence degree of R, we construct the algebra $S = \mathbf{Z}^+\{x_1, \ldots, x_k\}/J$, with J a T ideal. Say that S is an algebra of degree m and choose the monomials $N_1', \ldots, N_t' \in \mathbf{Z}^+\{x_1, \ldots, x_k\}$ as before relative to irreducible representations of degree $n \leqslant m$. Let $p_1, \ldots, p_t$ be given natural numbers and consider in S the ideal I generated by the elements $(\bar{N}_1')^{p_1}, (\bar{N}_2')^{p_2}, \ldots, (\bar{N}_t')^{p_t}$, ($\bar{N}_i'$ denotes the class of N_i' modulo J).

Consider finally the ring $\bar{S} = S/I$. $\bar{S}$ satisfies the condition of the theorem and it is therefore nilpotent of some degree q; furthermore if $R = \mathbf{Z}^+\{a_1, \ldots, a_k\}$ is an algebra satisfying the identities in the T ideal J such that the monomials N_i' computed in the a_i's are nilpotent of degree $\leqslant p_i$ respectively, then the canonical homomorphism

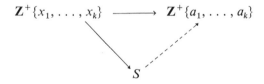

factors uniquely as above so R is nilpotent of degree $\leqslant q$.

Remark   In trying to find structure theorems for *PI* rings it has been found that if R is a *PI* ring then $L(R) = L_2(R)$ and $L_2(R)$ is nilpotent modulo $L_1(R)$. We have not proved any theorems about $L_1(R)$. In particular we can ask: Given a ring $R = L_1(R)$ what further conditions should R satisfy to assume that R is a *PI* algebra? Theorem **2.13**, just proved, gives us one.

For every k there should exist a number n and a function $\varphi: N \to N$ such that if $a_1, \ldots, a_k \in R$ and t is the degree of nilpotence of the monomials of degree $\leqslant n$ in the a_i's, then $\mathbf{Z}^+\{a_1, \ldots, a_k\}^{\varphi(t)} = 0$.

It is easy to see that this is a nontrivial condition; in fact if $S = \mathbf{Z}^+\{x_1, \ldots, x_k\}$ is the free algebra without 1, and n, t are two positive numbers, with $t > 1$, then considering the ideal I generated by the t powers of the monomials of degree $\leqslant n$, S/I is not nil. On the other hand, consider

$$J_m = I + \mathscr{S}^{(m)}, \qquad (\mathscr{S}^{(m)} = \sum_{i \geqslant m} S_i).$$

The ring $\oplus S/J_m$ is the sum of nilpotent ideals and it does not satisfy the condition above since S/I is not a nil ring.

In fact it is clear that this remark yields us one further condition for every PI ring.

Corollary 2.14   If R is a PI ring there exist two functions $\varphi: N \to N$, $\psi: N \to N$ such that if $a_1, \ldots, a_k \in R$ then for every $t \in N$ we have that every monomial in the a_i's of degree $s \geqslant \psi(t)$ is of the form

$$\sum f_i(a_1, \ldots, a_k) N_i{}^t q_i(a_1, \ldots, a_k)$$

where f_i, q_i are polynomials the a_i's, N_i is a monomial in the a_i's of degree $\leqslant \varphi(k)$, and the sum of the degrees of f_i, $N_i{}^t$, and q_i can be taken equal to s.

§3   The General Case

We ask if Theorem **2.10** can be generalized so that R is any algebra over a commutative ring A and "algebraic" is replaced by "integral."

As was mentioned before, we have only partial results in this direction that are interesting enough to be revealed. To further clarify what we will be able to prove, we state a conjecture. Fix a commutative ring B and natural numbers m, k.

Conjecture 3.1* There exist polynomials $M_1, \ldots, M_s \in \mathbf{Z}\{x_1, \ldots, x_k\}$ such that, for every algebra $R = A\{a_1, \ldots, a_k\}$ of degree $\leqslant m$ over a commutative B algebra A, considering the map $\lambda \colon \mathbf{Z}\{x_1, \ldots, x_k\} \to A\{a_1, \ldots, a_k\}$ given by $\lambda(x_i) = a_i$, we have that R is a finite A module if and only if the elements $\lambda(M_i)$ are integral over A.

The conjecture depends on the "parameters" B, m, k. Clearly when $B = \mathbf{Z}$ is the integers, we have the most general conjecture, since any commutative ring is a $\mathbf{Z}$ algebra. We will prove that the conjecture holds if B is a field of characteristic 0.

First we notice that, if the conjecture is always true, then the number of elements necessary to generate R as an A module can be estimated a priori using only knowledge of the degrees of the equations satisfied by the elements $\lambda(M_i)$ and also knowledge of the identities satisfied by R. To prove this we can pass to the generic situation.

Let $\mathscr{C}$ be the variety generated by R, and $n_1, \ldots, n_s$ the degrees of the monic polynomials satisfied by $\lambda(M_1), \ldots, \lambda(M_s)$. Consider the polynomial ring C over B in the variables

$$\xi_{11}, \xi_{12}, \ldots, \xi_{1n_1}, \quad \xi_{21}, \ldots, \xi_{2n_2}, \ldots, \xi_{s1}, \ldots, \xi_{sn_s}.$$

Consider the free algebra $S = C\{x_1, \ldots, x_k\}$. In S consider the ideal J generated by the polynomials $M_i^{n_i} + \xi_{i1} M_i^{n_i - 1} + \cdots + \xi_{in_i}$ where $i = 1, \ldots, s$ and by all the polynomial identities of R computed for all elements of S.

If one considers the algebra $S/J = \bar{S}$ one sees that it belongs to the variety $\mathscr{C}$ by construction. The construction of $\bar{S}$ depends on the following data: the variety $\mathscr{C}$, the number of variables k, the polynomials $M_1, \ldots, M_s$, and the numbers $n_1, \ldots, n_s$.

$\bar{S}$ is a C algebra, $\bar{S} = C\{\bar{x}_1, \ldots, \bar{x}_k\}$, calling $\bar{x}_i$ the class modulo J. The elements $\bar{M}_1, \ldots, \bar{M}_s \in \bar{S}$ satisfy, by construction, the polynomial equations

$$\bar{M}_i^{n_i} + \bar{\xi}_{i1} \bar{M}_i^{n_i - 1} + \cdots + \bar{\xi}_{in_i} = 0.$$

Going back to the algebra $R = A\{a_i, \ldots, a_k\}$ from which we started, we know that the elements $\lambda(M_i)$ satisfy polynomials

$$\lambda(M_i)^{n_i} + \beta_{i1}(M_i)^{n_i - 1} + \beta_{i2}\lambda(M_i)^{n_i - 2} + \cdots + \beta_{in_i} = 0.$$

---

* It is proved now in the addendum.

If we consider the B map $\tau\colon C \to A$, given by $\tau(\xi_{ij}) = \beta_{ij}$, we can construct a map

$$\tau'\colon C\{x_1, \ldots, x_k\} \to A\{a_1, \ldots, a_k\} = R$$

so that $\tau'(c) = \tau(c)$, $c \in C$, and $\tau'(x_i) = a_i$. We see that

$$\tau'(M_i^{n_i} + \xi_{i1}M_i^{n_i-1} + \cdots + \xi_{in_i}) = \lambda(M_i)^{n_i} + \beta_{i1}\lambda(M_i)^{n_i-1}$$
$$+ \cdots + \beta_{in_i} = 0.$$

Further, any polynomial identity of R computed in S gives an element of the kernel of τ' and so we see that $\tau'(J) = 0$. The map τ' factors as shown:

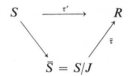

$$\bar{S} = S/J$$

If S/J is considered as a C algebra and R as an A algebra, we have $\bar{\tau}(cm) = \tau(c)\bar{\tau}(m)$, with $c \in C$, $m \in \bar{S}$. Through the map $\tau\colon C \to A$ we can extend the scalars of $\bar{S}$ and consider $\bar{S}_A = \bar{S} \otimes_C A$; we have an induced map

$$\bar{\tau}\colon \bar{S}_A = \bar{S} \otimes_C A \to R$$

where $\bar{\tau}(m \otimes a) = \bar{\tau}(m)a$. $\bar{\tau}$ is an A map and is surjective since $\bar{\tau}(\bar{x}_i \otimes 1) = a_i$.

Now let us draw the conclusions of all this construction.

If the polynomials $M_1, \ldots, M_s$ satisfy the property of **3.1** relative to B, m, k, where m is chosen in such a way that all algebras of $\mathscr{C}$ are of degree $\leqslant m$ (Chapter III, **2.4**), we must have, since $\bar{S} \in \mathscr{C}$ and the elements $\bar{M}_i \in \bar{S}$ are integral over C, that $\bar{S}$ is a finite C module, say generated by n elements. Since R is a homomorphic image of $\bar{S} \otimes_C A$ we see that R can also be generated by n elements. In fact one can say, a priori, which elements should be chosen, i.e., the images of the generaters of S over C. Conversely if S is finitely generated as a C module and this is true for all choices of $n_1, \ldots, n_s$ (and $\bar{S}$ depends on this choice), then the polynomials $M_1, \ldots, M_s$ satisfy the property of **3.1** for all algebras of $\mathscr{C}$. If we also let $\mathscr{C}$ vary, subject only to the restriction that deg $R \leqslant m$ when $R \in \mathscr{C}$ and we still have that the $\bar{S}$ relative to $\mathscr{C}$ is a finite C module, then the polynomials $M_1, \ldots, M_s$ satisfy the property of **3.1** relative to B, m, k.

At the end of this discussion we see that in fact we have proved the following useful reduction.

Reduction 3.2   Assume that the polynomials $M_1, \ldots, M_s \in \mathbf{Z}\{x_1, \ldots, x_k\}$ are such that for every algebra $R = A\{a_1, \ldots, a_k\}$ of degree $\leqslant m$ over a commutative B algebra A, *finitely generated over* B, considering the map

$$\lambda: \mathbf{Z}\{x_1, \ldots, x_k\} \to A\{a_1, \ldots, a_k\}$$

given by $\lambda(x_i) = a_i$, we have that R is a finite A module if and only if the elements $\lambda(M_i)$ are integral over A. Then the same statement is true upon dropping the hypothesis that A is finitely generated over B.

The use of this reduction is that, if B is chosen to be UJ, for instance $B = \mathbf{Z}$ or B is a field, we have that any finitely generated B algebra is a UJ ring so we have to study the problem in **3.1** only for A as a UJ ring. Therefore we will be able to apply the results that we already have proved on algebras over UJ rings and over noetherian rings.

The next reduction is, as has often been done, a reduction to the prime case.

We assume from now on that B is chosen to be a UJ integral domain (e.g., $B = \mathbf{Z}$ or B is a field).

We use **2.6** and **3.2** and immediately derive the following statement.

Reduction 3.3   The polynomials $M_1, \ldots, M_s$ satisfy the condition of **3.1** for all algebras of degree $\leqslant m$ if and only if they satisfy such condition for all prime algebras of degree $\leqslant m$.

Let us see to what point our reductions have led us.

We want to see whether some fixed polynomials $M_1, \ldots, M_s$ satisfy the condition of **3.1** for all prime algebras of degree $\leqslant m$ over finitely generated B algebras A. We fix an $n \leqslant m$, and a prime algebra $R = A\{a_1, \ldots, a_k\}$ of degree n, with A a finitely generated B algebra such that the polynomials M_i computed in the a_i's are integral over A. Let Q be the ring of fractions of R, Z the center of Q, and $\bar{Z}$ an algebraic closure of Z. Let

$$\psi : B\{x_1, \ldots, x_k\} \to (B[\xi_{ij_t}])_n$$

be the universal map in $n \times n$ matrices. If we choose a splitting map $Q \otimes_Z \bar{Z} \simeq (\bar{Z})_n$ we have a diagram

$$B\{x_1, \ldots, x_k\} \longrightarrow B\{\xi_1, \ldots, \xi_k\} \longrightarrow R$$

with vertical maps to $(B[\xi_{ij_t}])_n$ (via ψ), Q, and $Q \otimes_Z \bar{Z} \simeq (\bar{Z})_n$.

Let T be the ring generated by all coefficients of characteristic polynomials of elements of $B\{\xi_1, \ldots, \xi_k\}$ and T' the subring of T generated by the coefficients of the characteristic polynomials of the polynomials $M_1, \ldots, M_s$ computed in the elements $\xi_1, \ldots, \xi_k$.

We consider the ring

$$S = TB\{\xi_1, \ldots, \xi_k\} = T\{\xi_1, \ldots, \xi_k\}$$

and the subring $S' = T'\{\xi_1, \ldots, \xi_k\}$. We have that $S \subset (B[\xi_{ij_t}])_n$ and under the map $(B[\xi_{ij_t}])_n \to (\bar{Z})_n$, S is mapped in Q (cf. Chapter IV, §2). Now S' is a T' algebra, of degree n, generated by the elements $\xi_1, \ldots, \xi_k$. The polynomials $M_1, \ldots, M_s$ computed in the elements $\xi_1, \ldots, \xi_k$ are integral over T' since they satisfy their characteristic polynomials and the coefficients of such polynomials are by construction in T'. Therefore if $M_1, \ldots, M_s$ satisfy the conditions of 3.1 we have that S' is a finite T' module. Conversely assume S' is a finite T' module; we claim then that R is also a finite A module. Let $\bar{T}'$ be the image of T' in Z, the center of Q, and let D be the subring of Z generated by $\bar{T}'$ and $A \cdot 1 \subset Z$. Since T' is a finitely generated B algebra, D is a finitely generated A algebra. Let us call γ the map

$$\gamma : T'\{\xi_1, \ldots, \xi_k\} \to D\{a_1, \ldots, a_k\} \subseteq Q.$$

If $\bar{M}_i$ denotes the polynomial M_i computed in the ξ_j's and $\bar{\bar{M}}_i$ the same polynomial computed in the a_j's, we have $\gamma(\overline{M}_i) = \overline{\overline{M}}_i$. By hypothesis $\overline{\overline{M}}_i$ is integral over A, so the coefficients of the characteristic polynomial of $\overline{\overline{M}}_i \in Q$ are integral over A; now these coefficients are just the coefficients of the characteristic polynomial of $\overline{M}_i$ transformed through the map γ. It follows that the generators of T' are mapped via γ into integral elements over A. Therefore D is integral over A and, being finitely generated over A, it is a finite A module.

Furthermore, since S' is a finite T' module we have that $D\{a_1, \ldots, a_k\}$ is a finite D module; it follows that $D\{a_1, \ldots, a_k\}$ is a finite A module. Now A is a finitely generated B algebra, hence noetherian, and so it follows that

$$R = A\{a_1, \ldots, a_k\} \subseteq D\{a_1, \ldots, a_k\}$$

is also a finitely generated A module.

Having made this remark, we now have to study what condition the M_i's should satisfy to ensure that S' be a finite T' module. Then we will try to see whether such M_i's exist.

Since T' is a UJ ring, to be sure that $S' = T'\{\xi_1, \ldots, \xi_k\}$ is a finite A module it will be necessary and sufficient to show that S' is integral over T' (**2.7**), or even that S' is spanned, as T' module, by integral elements over T' (**2.5, 2.7**). Therefore it will be necessary and sufficient to prove that the elements of $B\{\xi_1, \ldots, \xi_k\}$ are integral over T'. Now an element of $B\{\xi_1, \ldots, \xi_k\}$ is integral over T' if and only if its characteristic polynomial has coefficients integral over T', so that S' is a finite T' module if and only if T is integral over T'.

Now we can come to the existence problem. We still fix $n \leqslant m$ and we try to find polynomials $M_1, \ldots, M_s$ such that, constructing T' and T as before, T is integral over T'. It is clear that if we can do this for all $n \leqslant m$, collecting all the polynomials that we obtain in this fashion, we will have a finite number of polynomials satisfying our condition **3.1**.

Lemma 3.4   Polynomials $M_1, \ldots, M_s$ with the preceding property exist if and only if T is a finitely generated B algebra.

Proof   Assume such polynomials exist. We know that T is integral over T', which is a UJ domain. The field of fractions of T is finitely generated over the field of fractions of B, therefore also over the field of fractions of T'. From this it follows that T is a finite T' modulo and a fortiori a finitely generated B algebra. Conversely, assume that T is a finitely generated B algebra. Since T is generated by the coefficients of the characteristic polynomials of the elements of $B\{\xi_1, \ldots, \xi_k\}$ we can find elements $c_1, \ldots, c_s$ such that the coefficients of the characteristic polynomials of $c_1, \ldots, c_s$ generate T. Now $c_1, \ldots, c_s$ can be considered as some polynomials $M_1, \ldots, M_s$ computed in the elements $\xi_1, \ldots, \xi_k$; therefore for such polynomials we have $T' = T$ and, a fortiori, T is integral over T'.

We have thus come to a question that we posed before (Chapter IV, §4). Before arriving at a positive result we make a final remark about **3.1**.

Lemma 3.5 T is a finitely generated B algebra if and only if $S = T\{\xi_1, \ldots, \xi_k\}$ is a finite T module.

Proof   We have already seen that if T is finitely generated then S is a finite T module. Assume that S is a finite T module and let $u_1, \ldots, u_s$ be linear generators of S over T. We will have $\xi_i = \sum_j t_{ij} u_j$, $u_i u_j = \sum_k t_{ijk} u_k$ for suitable elements t_{ij}, $t_{ijk} \in T$. Let $H = B[t_{ij}, t_{ijk}] \subset T$, and $L = \sum H u_i$. L is an H algebra since $t_{ijk} \in H$, and $L \supset B\{\xi_1, \ldots, \xi_k\}$ since $t_{ij} \in H$. Therefore if $u \in B\{\xi_1, \ldots, \xi_k\}$, then u is integral over H (since H is noetherian) and so the coefficients of its characteristic polynomial are integral over H. It follows that T is integral over H; again since the quotient field of T is finitely generated over the quotient field of H it follows, as H is a UJ ring, that T is of finite type over H and so a finitely generated B algebra.

As a consequence we see that **3.1** is equivalent to this apparently weaker statement.

Conjecture 3.6   If $R = A\{a_1, \ldots, a_k\}$ is any prime PI algebra of degree $\leqslant m$, integral over A where A is a B algebra, then R is a finite A module.

Proof   Conjecture **3.1** clearly implies this statement; conversely, if this is true, we can apply it to $S = T\{\xi_1, \ldots, \xi_k\}$ considered as a T algebra.

§4   Some Classical Invariant Theory

To be able to solve the problem set forth in the preceding paragraph we will have to treat a special case of the following problem.

Let G be a group acting on a finite dimensional vector space V over a field F. Let $S(V)$ be the symmetric algebra over V. The group G acts naturally as a group of automorphisms of the algebra $S(V)$. We consider $S(V)^G$, the invariant subalgebra under the G action; we ask whether this algebra is finitely generated. In any case, $S(V)^G$ is a graded subalgebra.

We start our investigations studying completely reducible G modules, with G a group.

Lemma 4.1   Let M be a completely reducible G module. Let $M^G = \{m \in M / gm = m, g \in G\}$ the invariant submodule.

(1)   There is a *unique* G map $\pi_M : M \to M^G$ which is a projection on M^G.

(2)   If $N \to M$ is a G submodule, then $\pi_M | N$ coincides with the unique projection $\pi_N : N \to N^G$.

Proof   (1)   This is clearly the same as to say that there is a unique submodule M_G of M such that $M_G \oplus M^G = M$. Let M_G be the sum of all irreducible submodules of M which are not contained in M^G. We claim that M_G satisfies our requirements.

First, since M is the sum of its irreducible submodules, we have $M = M_G + M^G$.

Second, consider $M_G \cap M^G$; this is a submodule of M_G which is completely reducible and so we have a projection map $\varphi : M_G \to M_G \cap M^G$. Every irreducible submodule Q of M_G is mapped under φ either in 0 or into an isomorphic submodule. Now, if $\varphi(Q)$ is isomorphic to Q, then since $\varphi(Q) \subseteq M^G$ we would have $Q \subseteq M^G$, a contradiction. Since M_G is the sum of its irreducible submodules, we have $\varphi(M_G) = 0$ and so $M_G \cap M^G = 0$. Finally if P is any submodule of M such that $P \cap M^G = 0$, then P is a sum of irreducible submodules not contained in M^G and therefore $P \subseteq M_G$.

(2)   Let P be a direct summand of N; then $M = N \oplus P$, and clearly $M^G = N^G \oplus P^G$. Consider the two unique projections given by part (1):

$$\pi_N \colon N \to N^G, \qquad \pi_P \colon P \to P^G.$$

Their direct sum $\pi_N \oplus \pi_P \colon N \oplus P \to N^G \oplus P^G$ gives a projection $M \to M^G$ and so by uniqueness we must have $\pi_N \oplus \pi_P = \pi_M$ which proves this second part.

Assume that G acts, as before, as a group of automorphisms of an integral domain A (not necessarily commutative). Assume that A is completely reducible as a G module. Let $J \colon A \to A^G$ be the unique projection map. J is called the Reynolds operator.

Lemma 4.2 (Reynolds identity)
  (1) $J(ab) = aJ(b)$ if $a \in A^G$.
  (2) $J(ab) = J(a)b$ if $b \in A^G$.

Proof   We prove (1). Consider aA; since $a \in A^G$ we see that aA is a G submodule and $(ab)^g = ab^g$. Therefore, since A is an integral domain, $aA^G = (aA)^G$. Clearly the map $ab \to aJ(b)$ is well defined and it is a projection $aA \to (aA)^G$; therefore by the previous lemma it coincides with J, and so $aJ(b) = J(ab)$. The proof of (2) is similar.

Before starting our next result we recall the convention that if $A = \oplus_{i=0}^{\infty} A_i$ is a graded algebra, we denote $A_+ = \oplus_{i=1}^{\infty} A_i$, and if $u \in A$ is a homogeneous element, we denote its degree by ∂u.

Lemma 4.3   Let $R = A[a_1, \ldots, a_k]$ be a finitely generated commutative algebra over a noetherian ring A. If $S \subseteq R$ is an A subalgebra and the elements a_i are integral over S, then S is a finitely generated A algebra.

Proof   Let

$$a_i^{n_i} + s_{i1}a_i^{n_i - 1} + \cdots + s_{in_i} = 0$$

with $i = 1, \ldots, k$, and $s_{ij} \in S$ be the integral dependence relations of the elements a_i over A. Let $T = A[s_{ij}] \subseteq S$. T is a finitely generated A algebra and the elements a_i are integral over T. Therefore R is a finite T module, S is a T submodule of R, A is noetherian and so T is also noetherian. Therefore S is a finitely generated T module. If $u_1, \ldots, u_m$ generate S over T, we clearly have $S = A[s_{ij}, u_t]$.

We come to our main theorem on invariants.

Theorem 4.4   Let G act on a finite-dimensional vector space V over a field F, and assume that $R = S(V)$ is a completely reducible G module.

(1) $U = R^G$ is a finitely generated F algebra.
(2)   If $T \subseteq U$ is a graded subalgebra and U_+ vanishes for every zero of the elements of T_+ then U is a finite T module and T is also a finitely generated F algebra.

Proof   (1)   Consider the ideal RU_+ of R. Since R is a noetherian ring, RU_+ is a finitely generated ideal: $RU_+ = \sum_{i=1}^k Ru_i$, $u_i \in U_+$. We can choose the u_i to be homogeneous since U_+ is a homogeneous subspace of R. We claim that $U = F[u_1, \ldots, u_k]$.
   Clearly $F[u_1, \ldots, u_k] \subset U$ and we must show the reverse inclusion. Clearly $U_0 = F \subset F[u_1, \ldots, u_k]$ and therefore we can use an induction argument and assume that $U_i \subset F[u_1, \ldots, u_k]$ for $i < n$. Let $u \in U_n$; since $u \in RU_+$ we have $u = \sum_{i=1}^k r_i u_i$, with $r_i \in R$. We can clearly choose the r_i's to be homogeneous and such that $\partial r_i + \partial u_i = \partial u = n$. In particular, since $\partial u_i > 0$, then $\partial r_i < n$. We apply the Reynolds operator and obtain $u = J(u) = \sum_{i=1}^k J(r_i)u_i$, by **4.2**. Now J preserves degrees, since R_n is a G submodule and $J: R_n \to R_n{}^G$ (**4.1**(2)). Therefore $J(r_i) \in U_{\partial r_i}$ and, as $\partial r_i < n$, then $U_{\partial r_i} \subset F[u_1, \ldots, u_k]$. This finally shows that

$$F[u_1, \ldots, u_k] \supset U_n$$

and so by induction $U = F[u_1, \ldots, u_k]$.
   (2) $S(V)$ can be considered as a polynomial ring and we apply to it the Hilbert Nullstellensatz; if U_+ vanishes on the zeroes of T_+ we must have $U_+{}^n \subseteq RT_+$ for a suitable exponent n.
   Let $u \in U_+{}^n$; we have $u = \sum r_j t_j$, with $r_j \in R$, $t_j \in T_+$. Applying the Reynolds operator we have $u = J(u) = \sum J(r_j)t_j$ since $t_j \in U$; therefore $U_+{}^n \subseteq T_+U$.

Next we claim that there is an index s such that $U_j \subseteq U_+^n$ if $j \geqslant s$. In fact let $u_1, \ldots, u_k \in U_+$ be generators of U over F as in (1), and assume a number m is chosen so that $\partial u_i \leqslant m$; put $s = nm$. Let $u \in U_j$, $u = \sum \alpha_{i_1 \ldots i_k} u_1^{i_1} \ldots u_k^{i_k}$ with $\sum_{t=1}^{k} i_t \partial u_t = j$. Since $\partial u_t \leqslant m$ we have $nm \leqslant j \leqslant (\sum_{t=1}^{k} i_t) m$ and therefore $\sum_{t=1}^{k} i_t \geqslant n$, and $u \in U_+^n$.

At this point we have proved that $UT_+ \supseteq U_j$ if $j \geqslant s$.

Now let $1 = c_0, c_1, \ldots, c_r$ be homogeneous linear generators for $\bigoplus_{i=0}^{s-1} U_i$. We claim that $U = \sum c_i T$. We work again by induction. Clearly $\bigoplus_{i=0}^{s-1} U_i \subseteq \sum c_i T$ since the c_i's generate this subspace over F. Assume then that we have proved $U_j \subseteq \sum c_i T$ for $j < n$ and let $n \geqslant s$. If $u \in U_n$, since $n \geqslant s$, we have $u \in UT_+$ and so $u = \sum u_i t_i$, with $u_i \in U$, $t_i \in T_+$ and we can choose all the elements to be homogeneous. Therefore $\partial u_i < n$ and by induction $u_i \in \sum c_i T$, so $u \in \sum c_i T$. To prove now that T is a finitely generated algebra it suffices to apply **4.3**.

Having developed the theory so far, we must give a sufficient criterion for $S(V)$ to be a fully reducible G module.

For algebraic groups G and algebraic representations $G \to GL(V)$, there is an extensive theory that is really good only in characteristic zero.

We do not need the full theory for our purposes and we will content ourselves with the following quite elementary result.

Proposition 4.5   Let V be a finite dimensional vector space over Q, the field of rational numbers, and let (v, w) be a nondegenerate symmetric bilinear form with $(v, v) > 0$ if $v \neq 0$ (we say a scalar product). Let $*$ denote the adjunction map $*: \mathrm{End}(V) \to \mathrm{End}(V)$ induced by the form $(\tau(v), w) = (v, \tau^*(w))$. If $A \subseteq \mathrm{End}(V)$ is a set of linear transformations such that $A = A^*$, then the induced A module structure on $T(V)$ (tensor algebra) and on $S(V)$ (symmetric algebra) is completely reducible.

Proof   Since $S(V)$ is a quotient module of $T(V)$ it is enough to prove that $T(V)$ is completely reducible. Since $T(V) = \bigoplus T_n(V)$ where $T_n(V) = V \otimes V \otimes \ldots \otimes V$ (n times), then $T_n(V)$ is invariant under A, and $a \in A$ induces on $T_n(V)$ the map $a_n : v_1 \otimes v_2 \otimes \ldots \otimes v_n \to av_1 \otimes av_2 \otimes \ldots \otimes av_n$. So it is enough to show that $T_n(V)$ is completely reducible. We define a bilinear form on $T_n(V)$ in the following way:

$$(v_1 \otimes v_2 \otimes \ldots \otimes v_n, w_1 \otimes w_2 \otimes \ldots \otimes w_n) = \prod_{i=1}^{n} (v_i, w_i).$$

It is a simple matter to show that it is nondegenerate and if $0 \neq w \in T_n(V)$, then $(w, w) > 0$.

We see that

$$(a_n(v_1 \otimes v_2 \otimes \ldots \otimes v_n), w_1 \otimes w_2 \otimes \ldots \otimes w_n) = \prod (av_i, w_i)$$

$$= \prod (v_i, a^*w_i) = (v_1 \otimes v_2 \otimes \ldots \otimes v_n, (a^*)_n(w_1 \otimes w_2 \otimes \ldots \otimes w_n)),$$

and therefore $(a_n)^* = (a^*)_n$. Therefore, if $A_n = \{a_n | a \in A\}$ then $A_n{}^* = A_n$. Now we show that in general if B is a set of operators over a finite dimensional vector space W over Q with a scalar product, we have that W is completely reducible under B if $B = B^*$; this will finish the proof of our proposition.

Let $M \subset W$ be a B submodule. Consider $M^\perp = \{w \in W | (w, m) = 0$ if $m \in M\}$; since the bilinear form is a scalar product we have $M \oplus M^\perp = W$. We claim that $M^\perp$ is a B submodule; in fact if $b \in B$, $v \in M^\perp$, and $m \in M$, we have $(bv, m) = (v, b^*m) = 0$ since $b^* \in B$ and so $b^*m \in M$ (since M is B-invariant). Therefore $bv \in M^\perp$ and the proposition is proved.

§5   The Main Theorem for $B = Q$

We go back to the terminology and the problem of §3. We take $Q = B$ and we consider the algebra $Q\{\xi_1, \ldots, \xi_k\}$, with $\xi_s = \sum_{ij=1}^{n} \xi_{sij} e_{ij}$ and $s = 1, \ldots, k$, a generic matrix in the ring $(Q[\xi_{sij}])_n$. $T \subset Q[\xi_{sij}]$ is the ring generated by the coefficients of the characteristic polynomials of the elements of $Q\{\xi_1, \ldots, \xi_k\}$. Our aim is to prove that T is a finitely generated Q algebra; this will prove the validity of 3.1 for $B = Q$.

We use the results of the previous section.

Lemma 5.1
(1) T is generated by the traces of the elements of $Q\{\xi_1, \ldots, \xi_k\}$.
(2) T is a graded subalgebra of $Q[\xi_{sij}]$.

Proof   (1)   Let $a \in Q\{\xi_1, \ldots, \xi_k\}$, and $\alpha_1, \ldots, \alpha_n$ be the characteristic roots of a computed with their multiplicities. The coefficients of the characteristic polynomial of a are, up to sign, the elementary symmetric functions $\sigma_i(\alpha_1, \ldots, \alpha_n)$, with $i = 1, \ldots, n$, of $\alpha_1, \ldots, \alpha_n$. Consider now the symmetric functions

$$\varphi_k(\alpha_1, \ldots, \alpha_n) = \sum_{i=1}^{n} \alpha_i^k, \qquad k = 1, \ldots, n.$$

One can express the functions σ_i as polynomials, with rational coefficients in the functions φ_k ([27]). Now $\varphi_k(\alpha_1, \ldots, \alpha_k) = \text{Tr}(a^k)$ and so (1) is proved.

(2) It is enough to remark that $Q\{\xi_1, \ldots, \xi_n\}$ is a graded subalgebra of $(Q[\xi_{sij}])_n$ and Tr is a linear map preserving degrees.

Let us now consider the action of $G = \text{Aut}(Q)_n$ on $Q[\xi_{sij}]$ and the invariant ring $U = Q[\xi_{sij}]^G$ (Chapter IV, §3). We have $T \subset U$ and we wish to show that we can apply Theorem **4.4** and deduce that T is a finitely generated Q algebra.

First we show that $Q[\xi_{sij}]$ is completely reducible under G and this will show that U is finitely generated.

The ring $Q[\xi_{sij}]$ can be thought of as $S(V)$ with $V = \sum \xi_{sij}Q$. We can identify V with the dual of $W = \oplus (Q)_n$ (k times) in the following way. If e_{sij} denotes the vector $(0, 0, \ldots, e_{ij}, 0, \ldots, 0)$ where e_{ij} is in the sth place in this k-tuple, we consider

$$\langle \xi_{sij}, e_{ths} \rangle = \delta_{st}\delta_{ih}\delta_{js}.$$

Under this identification, in fact, we have: if $\rho: Q\{x_1, \ldots, x_k\} \to (Q)_n$ is a morphism, this corresponds to an element $(\rho(x_1), \ldots, \rho(x_k))$ of W and the converse. If $\bar{\rho}: S(V) \to Q$ is the associated map so that

is commutative then $\bar{\rho}: S(V) \to Q$ is induced by a linear map $\bar{\rho}_V: V \to Q$. This linear map is identified with the element $(\rho(x_1), \ldots, \rho(x_k))$ of $W = V^*$.

The group G acts naturally coordinatewise on $W = \oplus (Q)_n$ (k times) and it is induced as inner automorphism by $GL(n, Q)$; i.e., if $A \in GL(n, Q)$ and $w = (B_1, \ldots, B_k) \in W$, we have the automorphism γ_A induced by A, where $\gamma_A(w) = (AB_1A^{-1}, \ldots, AB_kA^{-1})$; similarly A induces an inner automorphism $\bar{\gamma}_A$ of $(S(V))_n$.

We consider the transpose γ_A^* of γ_A on V (identified with W) and we want to verify that the induced map $\bar{\gamma}_A^*: S(V) \to S(V)$ is such that the diagram

$$
\begin{array}{ccc}
Q\{x_1, \ldots, x_k\} & \longrightarrow & (S(V))_n \\
 & \searrow_{\bar{\gamma}_A \varphi} & \downarrow_{(\bar{\gamma}_A^*)_n} \\
 & & (S(V))_n
\end{array}
$$

is commutative. In fact it is enough to check that, composing with any map $\rho_n: (S(V))_n \to (Q)_n$, the diagram

$$
\begin{array}{ccccc}
Q\{x_1, \ldots, x_k\} & \xrightarrow{\varphi} & (S(V))_n & \xrightarrow{\rho_n} & (Q)_n \\
 & \searrow_{\gamma_A \cdot \varphi} & & & \downarrow_{\gamma_A} \\
 & & (S(V))_n & \xrightarrow{\rho_n} & (Q)_n
\end{array}
$$

is commutative, since $\rho_n(\bar{\gamma}_*)_n = \gamma_A \rho_n$. This last diagram is commutative by Lemma **3.2**, Chapter IV.

Next we want to prove that V has a positive definite form satisfying the property of **4.5** relative to G. Since V is the dual of W and the action of G on V is the contragradient action, it is enough to show that W satisfies the condition of **4.5**.

Now on W consider the following bilinear form:

$$((A_1, \ldots, A_k), (B_1, \ldots, B_k)) = \sum \text{Tr}(A_i B_i^t).$$

It is clearly symmetric and positive definite since $\text{Tr}(AA^t) = \sum a_{ij}^2 > 0$ if $A = (a_{ij}) \neq 0$. Now if $A \in GL(n, Q)$, we have

$$
\begin{aligned}
((AA_1A^{-1}, AA_2A^{-1}, \ldots, AA_kA^{-1}), (B_1, B_2, \ldots, B_k) &= \text{Tr}(AA_iA^{-1}B_i^t) \\
&= \text{Tr}(A_iA^{-1}B_i^tA) = \text{Tr}(A_i(A^tB_i(A^t)^{-1})^t) \\
&= ((A_1, \ldots, A_k), (A^tB_1(A^t)^{-1}, \ldots, A^tB_k(A^t)^{-1})),
\end{aligned}
$$

so under the * map induced by the form we see that $(\gamma_A)^* = \gamma_A t$ and the condition is verified.

Finally we must show that $T \subseteq U$ satisfies the property of **4.4(2)**.

Therefore let $\varphi: Q[\xi_{sij}] \to K$ be a map such that $\varphi(T_+) = 0$. φ induces a map $\bar{\varphi}: Q\{\xi_1, \ldots, \xi_k\} \to (K)_n$ such that the diagram

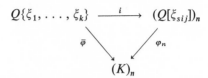

is commutative. Since $\varphi(T_+) = 0$ we have $\mathrm{Tr}(\bar{\varphi}(a)) = 0$ for every $a \in Q\{\xi_1, \ldots, \xi_k\}_+$; therefore the elements of the ring $\bar{\varphi}(Q\{\xi_1, \ldots, \xi_k\}_+)$ are all nilpotent and so can be brought to upper triangular form by an inner automorphism γ of $(K)_n$.

The map $\gamma\bar{\rho}: Q\{\xi_1, \ldots, \xi_k\} \to (K)_n$ gives rise by universality to a new map $\tilde{\varphi}: Q[\xi_{sij}] \to K$ such that $\tilde{\varphi}_n i = \gamma\varphi$. Since U is the ring of absolute invariants (Chapter IV, **3.9**), we have $\tilde{\varphi}|_U = \varphi|_U$. Let $u \in U$ be homogeneous of degree $m > 0$, $u = u(\xi_{sij})$. If $\tilde{\varphi}(\xi_{sij}) = \alpha_{sij}$ we have $\alpha_{sij} = 0$ if $i \leqslant j$ and $\varphi(u) = u(\alpha_{sij})$.

Next let $\tau \neq 0$, $\tau \in K$, and apply the inner automorphism of $(K)_n$ induced by the matrix

$$\begin{pmatrix} \tau & & & & \\ & 1 & & 0 & \\ & & 1 & & \\ & & & \ddots & \\ 0 & & & & 1 \end{pmatrix}$$

This automorphism transforms the matrix $\tilde{\varphi}(\xi_t) = \sum \alpha_{t,ij}e_{ij}$ into the matrix $\sum \alpha'_{t,ij}e_{ij}$ with $\alpha'_{t,ij} = \alpha_{t,ij}$ if $i > 1$ and $\alpha'_{t,1j} = \tau\alpha_{t,1j}$. Again using the fact that u is an absolute invariant, we see that $\varphi(u) = u(\alpha_{tij}) = u(\alpha'_{tij})$ and therefore, since τ is arbitrary, $u(\alpha_{tij})$ does not depend on the elements $\alpha_{t,1j}$.

Next we apply the inner automorphism

$$\begin{pmatrix} 1 & & & & \\ & \tau & & 0 & \\ & & 1 & & \\ & & & \ddots & \\ 0 & & & & 1 \end{pmatrix}$$

and in the same way verify that u does not depend on the $\alpha_{t_2 j}$.

Inductively we see that u does not depend on any of the α_{tij} and so, being homogeneous of positive degree, $\varphi(u) = 0$. This shows that the second condition of **4.4** is satisfied and so the theorem is finally proved.

Addendum to Chapter VI

After writing Chapter VI it was pointed out to me by Prof. Amitsur that the conjectures stated there had been proved by the Russian mathematician A. I. Širšov in the following papers:

On certain non associative nil rings and algebraic algebras, Mat. Sb. N.S. 41(83) (1957), 381–394.

On rings with identity relations (Russian), Mat. Sb. N.S. 43(85) (1957), 277–283.

On the Levitzki problem (Russian), Dokl. Akad. Nauk 120 (1958), 41–42.

Širšov's methods are essentially combinatorial. In the meantime, confronting his methods with some of the results of Chapter VI. I realized that Conjecture 3.1 is an easy consequence of 2.14.

This is a quite interesting argument since it reduces a problem on finite generation to a problem about nilpotent algebras. The proof of 2.14, on the other hand, is based only on 2.11 and 2.13. Therefore, the general conjecture is proved in an extremely simple way by structure theory.

Here I want to give an improvement of 2.9 and 2.13, and deduce from these the proof of 3.1 via 2.14 in a precise form.

Proposition 1   Let $a_1, \ldots, a_k \in (K)_n$ be $n \times n$ matrices over a field K. If the subalgebra generated by $a_1, \ldots, a_k$ over K is $(K)_n$ then the monomials of degree $\leqslant n^2$ in $a_1, \ldots, a_k$ span $(K)_n$ over K.

Proof   Let V_i be the space spanned, over K, by the monomials of degree $\leqslant i$ in the elements $a_1, \ldots, a_k$. I claim that, if for some i we have $V_i = V_{i+1}$, then V_i is the subalgebra generated by the elements $a_1, \ldots a_k$.

In fact if $V_i = V_{i+1}$ this means that $a_j V_i \subseteq V_i$ for every a_j. From this, inductively, we have that $V_k V_i \subseteq V_i$ and the claim follows. Now in our case we know that $a_1, \ldots, a_k$ generate $(K)_n$. Let i_o be the minimum index such that $V_{i_o} = V_{i_o+1} = (K)_n$. Since for every $j < i_o$ we have $\dim_K V_j < \dim_K V_{j+1}$ we must have $i_o \leqslant n^2$. This proves the proposition.

Proposition 2   If $R = \{a_1, \ldots a_k\}$ is a PI algebra of degree $< m$ generated by the elements $a_1, \ldots, a_k$ and if the monomials in $a_1, \cdots, a_k$ of degree $\leqslant m^2$ are nilpotent, then R is nilpotent.

Proof   Let P be a prime ideal of R, passing to R/P we can assume that R is prime we have to show that $R = 0$.

Assume $R \neq 0$, let $R \to (K)_h$ be a central injection ($0 < h \leqslant m$). The elements $a_1, \ldots, a_k$ are thus considered as matrices in $(K)_h$ and they generate $(K)_h$. Now from Proposition 1 the monomials in $a_1, \ldots, a_k$ of degree $\leqslant h^2$ span $(K)_h$ over K. This is a contradiction since such monomials are nilpotent.

Theorem 3   Let $R = \Lambda \{a_1, a_2, \ldots, a_k\}$ be a PI algebra. Assume that the free algebra $Z\{x_1, \ldots x_k\}$ modulo the ideal J of stable identities satisfied by R has degree $\leqslant m$ and that the monomials in $a_1, \ldots, a_k$ of degree $\leqslant m^2$ are integral over Λ, then R is a finite Λ module.

Proof   Let $S = Z\{\bar{x}_1, \ldots, \bar{x}_k\} = Z\{x_1, \ldots x_k\}/J$ be the free algebra modulo the ideal J of stable identities of R. Since J is a homogeneous ideal, S is a graded algebra. Let $T_{(t)}$ be the ideal of S generated by the elements M^t, M varying between the monomials in $\bar{x}_1, \ldots, \bar{x}_k$ of degree d with $0 < d \leqslant m^2$. If S^+ denotes the positive part of S then S^+/T is nilpotent by Proposition 2. Call s_t the degree of nilpotence of S^+/T.

If N is any monomial in $\bar{x}_1, \ldots, \bar{x}_k$ of degree $\geqslant s_t$ we must have $N \in T$, i.e., $N = \sum_i q_i M_i r_i$, $q_i, r_i \in S$ and $0 < \deg M_i \leqslant m^2$. Since S is a graded ring we can assume that $\deg q_i + t \deg M_i + \deg r_i = \deg N$.

Assume now that the monomials of degree $\leqslant m^2$ in R are integral and satisfy monic polynomials of degree $\leqslant t$.

Let s_t be as before. I claim that R is spanned by the monomials of degree $< s_t$. We proceed by induction, call U_l the submodule of R generated by the monomials in $a_1, \ldots, a_k$ of degree $\leqslant l$. We must show that $R = U_{s_t-1}$. Assume by induction that for some l we have $U_l \subseteq U_{s_t-1}$ then we prove that $U_{l+1} \subseteq U_{s_t-1}$, this will clearly finish the proof of the theorem.

Now if $l + 1 < s_t$, the claim is clear, otherwise assume $l + 1 \geqslant s_t$. Let N be a monomial of degree $l + 1$ in $\bar{x}_1, \ldots, \bar{x}_k$, we must show that this monomial, computed in $a_1, \ldots, a_k$, lies in $U_l \subseteq U_{s_t-1}$.

In the free algebra S we have the relation $N = \sum_i q_i M_i^t r_i$, $\deg M_i \leqslant m^2$ and $q_i M_i^t r_i$ homogeneous of degree $l + 1$. Now the monomial M_i computed in $a_1, \ldots, a_k$ satisfies a monic polynomial of degree t with coefficients in Λ; therefore we see that M_i^t computed in $a_1, \ldots a_k$ lies in the submodule spanned over Λ by the monomials of degree $< \deg M_i^t$. This in turn implies that N lies in the submodule spanned by the monomials of degree $< l + 1$, i.e., in U_l as it was announced.

This proves the theorem in full.

Remark 1   If R satisfies a multilinear identity of degree h then the degree of S is $\leqslant h/2$.

Remark 2   Let $\varphi(m)$ be the minimum number such that: if $a_1, \ldots, a_s$ are $h \times h$ matrices over a field K, $h \leqslant m$, and the monomials of degree $\leqslant \varphi(m)$ are nilpotent then the subalgebra over K generated by $a_1, \ldots, a_s$ is not $(K)_h$. We know that $\varphi(m) \leqslant m^2$.

$\varphi(m)$ is also the minimum number that we can substitute to m^2 in the statement of Theorem 3, without loosing the validity of the same theorem.

Therefore, any improvement to the estimate $\varphi(m) \leqslant m^2$, will be an improvement to Theorem 3.

Chapter VII

INTRINSIC CHARACTERIZATION OF AZUMAYA ALGEBRAS

§1 Localization

Let R be a ring with 1 and $M \subseteq R$ a subset.

Definition 1.1   A universal M inverting map is a map $\lambda: R \to S$ such that:

(1) $\lambda(M)$ consists of invertible elements.
(2) If $\mu: R \to T$ is a map such that $\mu(M)$ consists of invertible elements then there is a unique map $\bar{\mu}: S \to T$ such that $\mu = \bar{\mu}\lambda$.

It is clear that if a universal M inverting map exists then it is unique up to a unique isomorphism.

The existence of such a map is trivially given by generators and relations as follows:

Let $\mathbf{Z}\{x_m\}_{m \in M}$ be the free algebra in the variables x_m indexed by the set M (with $\mathbf{Z}$ the integers). Consider the free product $U = R * \mathbf{Z}\{x_m\}$ and in U the ideal I generated by the elements $x_m m - 1$, $m x_m - 1$. Let $S = U/I$ and

$$\lambda: R \to R * \mathbf{Z}\{x_m\} \to U/I = S$$

be the canonical map; it is clear that λ satisfies the conditions of **1.1**.

Let $M_1 \subseteq M_2$ be two multiplicative sets and $\lambda_1: R \to S_1$, $\lambda_2: R \to S_2$ be the two universal inverting maps for M_1, M_2 respectively. It is clear that, by the universal property, there is a unique map $\bar{\lambda}_2: S_1 \to S_2$ such that $\lambda_2 = \bar{\lambda}_2\lambda_1$.

In general it is quite hard to say anything about this universal map, so that it actually turns out to be quite useless except in the special important cases that we want to discuss.

First we assume that M is closed under multiplication, then we set the following.

Definition 1.2   The universal M inverting map $\lambda: R \to S$ is called a right (respectively left) localization at M if

$$S = \{\lambda(a)\lambda(m)^{-1}|a \in R, m \in M\}$$

(respectively, $S = \{\lambda(m)^{-1}\lambda(a)|a \in R, m \in M\}$).

Theorem 1.3

(1) $\lambda: R \to S$ is a right localization at M if and only if the set $U = \{\lambda(a)\lambda(m)^{-1}|a \in R, m \in M\}$ is a subring.

(2) U is certainly a subring if the following conditions are satisfied.

   (a)   Given $a \in R$, $m \in M$, there exist elements $n \in M$, $b \in R$ such that $mb = an$ (Öre condition).

   (b)   If $ab \in M$ and $a \in M$ then $b \in M$.

Proof   (1) If U is a subring it is clear that the map $\lambda: R \to U$ satisfies conditions (1) and (2) and so by uniqueness $U = S$.

(2) Assume that the condition is satisfied. Let $\alpha = \lambda(a)\lambda(m)^{-1}$, $\beta = \lambda(c)\lambda(p)^{-1} \in U$; we must show that $\alpha - \beta$ and $\alpha\beta \in U$. Now there exist elements $b \in R$, $n \in M$ such that $cn = mb$ so that $\lambda(c)\lambda(n) = \lambda(m)\lambda(b)$ and therefore $\lambda(m)^{-1}\lambda(c) = \lambda(b)\lambda(n)^{-1}$. Thus we deduce

$$\alpha\beta = \lambda(a)\lambda(m)^{-1}\lambda(c)\lambda(p)^{-1} = \lambda(a)\lambda(b)\lambda(n)^{-1}\lambda(p)^{-1} = \lambda(ab)\lambda(pn)^{-1},$$

as $pn \in M$ since it is a multiplicative subset.

As for addition, there exist elements $u \in R$, $v \in M$ such that $mu = pv$.

By condition (b) we have $u \in M$ and so

$$\alpha - \beta = \lambda(a)\lambda(u)\lambda(u)^{-1}\lambda(m)^{-1} - \lambda(c)\lambda(v)\lambda(v)^{-1}\lambda(p)^{-1}$$
$$= \lambda(au)\lambda(mu)^{-1} - \lambda(cv)\lambda(pv)^{-1} = \lambda(au - cv)\lambda(mu)^{-1} \in U.$$

Assume that $\lambda \colon R \to S$ is a left localization at M; we write $S = R_M$. We have the following important property for the ideal structure of R_M.

Proposition 1.4   If $I \subseteq R_M$ is a right ideal then $I = \lambda(\lambda^{-1}(I))R_M$.

Proof   Clearly $I \supseteq \lambda(\lambda^{-1}(I))R_M$. Let $\alpha \in I$, $\alpha = \lambda(a)\lambda(m)^{-1}$, $a \in R$, $m \in M$. Then $\alpha\lambda(m) = \lambda(a) \in I$ so that $a \in \lambda^{-1}(I)$ and, therefore, $\alpha = \lambda(a)\lambda(m)^{-1} \in \lambda(\lambda^{-1}(I))R_M$.

This proposition implies a useful corollary.

Corollary 1.5
(1)   If $I_1 \neq I_2$ are right ideals of R_M then $\lambda^{-1}(I_1) \neq \lambda^{-1}(I_2)$.
(2)   Let $M_1 \subseteq M_2$ be multiplicative sets and let $\lambda_1 \colon R \to S_1$, $\lambda_2 \colon R \to S_2$ be the universal inverting maps for M_1, M_2 and $\bar{\lambda}_2 \colon S_1 \to S_2$ the unique map such that $\lambda_2 = \bar{\lambda}_2\lambda_1$. Assume that $S_1 = R_{M_1}$ is a right localization and that λ_2 is injective. Then $\bar{\lambda}_2$ is injective.

Proof   (1)   The proof is an immediate consequence of **1.4.**
(2)   Let $I = \mathrm{Ker}\,\bar{\lambda}_2$, then $\lambda^{-1}(I) \cap \mathrm{Ker}\,\lambda_2 = 0$, so that $\lambda^{-1}(I) = 0$, and by **1.4,** $I = 0$.

We will use this corollary in particular when R is a prime PI ring and M is a multiplicative set of regular elements such that the universal M inverting map is a right localization; by the previous corollary it follows that R_M is contained in the total ring of quotients of R.
    The way in which polynomial identities intervene is to make sure that in some suitable cases the conditions of **1.3**(2) are satisfied; here is the main theorem.

Theorem 1.6   Let R be a PI ring satisfying a multilinear identity

$$\gamma(x_1, \ldots, x_n) = x_1\psi(x_2, \ldots, x_n) + \varphi(x_1, \ldots, x_n),$$

where in $\varphi(x_1, \ldots, x_n)$ we collect all monomials which do not start with x_1. Let I be a two-sided ideal of R.

(1)   If R/I does not satisfy the identity $\psi(x_2, \ldots, x_n)$ and I is a prime ideal, then $M = \{r \in R | \bar{r} \in R/I$ is regular$\}$ satisfies the conditions of **1.3**(2) (on the right).

(2)   If R/J does not satisfy the identity $\psi(x_2, \ldots, x_n)$ for every proper ideal $J \supseteq I$, then $M = \{r \in R | \bar{r} = 1$ in $R/I\}$ satisfies the condition of **1.3**(2) (on the right).

Proof   In both cases it is easy to see that M is a multiplicative set and if $uv \in M$ and $u \in M$ then $v \in M$ (similarly, if $uv \in M$ and $v \in M$ then $u \in M$). For case (2) the proof is trivial. For case (1) the proof follows from the analogous property on the set of regular elements of a prime *PI* ring; this in turn is an immediate consequence of the existence of a ring of quotients for such rings. In both cases we have to verify the Öre condition.

(1)   Let $m \in M$, $a \in R$; we have for $r_2, \ldots, r_n \in R$ that

$$0 = \gamma(a, mr_2, \ldots, mr_n) = a\psi(mr_2, \ldots, mr_n) + mt,$$

where $t \in R$ is a suitable element depending on $a, r_2, \ldots, r_n$. Therefore, for all choices of $a, r_2, \ldots, r_n$ the two-sided ideal generated by $\psi(mr_2, \ldots, mr_n)$ is contained in mR; when $\bar{R} = R/I$ is a prime ring and $\bar{m}$ is regular in $\bar{R}$, then $\bar{R}$ and $\bar{m}\bar{R}$ have the same quotient ring and hence the same identities [Chapter II, **5.7**(4)]. Now $\psi(x_2, \ldots, x_n)$ is not an identity of $\bar{R}$ so there are elements $s_2, \ldots, s_n$ such that $\psi(ms_2, \ldots, ms_n) \notin I$. If J is the two-sided ideal generated by $\psi(ms_2, \ldots, ms_n)$ we have that $J \not\subseteq I$ and we can find an element $n \in J$, regular modulo I (Chapter II, **5.5**), so that $n \in M$.

We have thus proved for every $a \in R$ that $an \in mR$, and so the Öre condition is satisfied in a strong form (fixing $m \in M$ we find an $n \in M$ which is good for all $a \in R$).

(2)   We see from the hypothesis that the ideal generated by the elements $\psi(b_2, \ldots, b_n)$ where $b_i \in R/I$ is the whole R/I (otherwise we would find a nonzero quotient of R satisfying $\psi(x_2, \ldots, x_n)$). Therefore let $a_i, c_i, r_j^{(i)} \in R$, with $j = 2, \ldots, n$, $i = 1, \ldots, s$, be elements such that

$$u = \sum a_i \psi(r_2^{(i)}, \ldots, r_n^{(i)}) c_i \equiv 1 \qquad \text{mod } I.$$

Let $a \in R$, $m \in M$; we see that

$$0 = \gamma(aa_i, mr_2^{(i)}, \ldots, mr_n^{(i)}) = aa_i\psi(mr_2^{(i)}, \ldots, mr_n^{(i)}) + mt,$$

with $t \in R$ a suitable element. Therefore,

$$a \sum a_i\psi(mr_2^{(i)}, \ldots, mr_n^{(i)})c_i = mb \qquad \text{for some} \quad b \in R.$$

Now if we set $n = \sum a_i\psi(mr_2^{(i)}, \ldots, mr_n^{(i)})c_i$, we see that since $\bar{m} = 1$ mod I,

$$\bar{n} = \sum \bar{a}_i\psi(\bar{m}\bar{r}_2^{(i)}, \ldots, \bar{m}\bar{r}_n^{(i)})\bar{c}_i = \sum \bar{a}_i\psi(\bar{r}_2^{(i)}, \ldots, \bar{r}_n^{(i)})\bar{c}_i = \bar{u} = 1.$$

Therefore $n \in M$ and also in this case the Öre condition is satisfied.

Assume that P is a prime ideal of R and consider

$$S = \{r \in R | r \text{ is regular modulo } P\}.$$

S is a multiplicative set. Assume that the conditions of **1.3**(2) are satisfied for S, in particular that the S inverting map is a right localization. Then instead of R_S we write R_P, as in commutative algebra.

Assume under the previous conditions that R is a *PI* ring. Let $\lambda: R \to R_P$ be the universal map.

Definition 1.7   We say that a ring R is a local ring if it has a unique maximal ideal M.

Remark   Some authors assume also that R/M should be a division ring, but this is not a useful definition for our purposes.

Proposition 1.8 R_P is local with a unique maximal ideal $\lambda(P)R_P$.

Proof   First let us check that $\lambda(P)R_P$ is a two-sided ideal of R_P and that $\lambda^{-1}(\lambda(P)R_P) = P$. Let $u \in P$ and $\lambda(a)\lambda(r)^{-1} \in R_P$. By the Öre condition there is an $s \in S$ and $v \in R$ such that $us = rv$. Passing modulo P we have $\bar{u} = 0$ so $0 = \bar{r}\bar{v}$; since r is regular modulo P we have $\bar{v} = 0$ so that $v \in P$. Now $\lambda(a)\lambda(r)^{-1}\lambda(u) = \lambda(a)\lambda(v)\lambda(s)^{-1} = \lambda(av)\lambda(s)^{-1} \in \lambda(P)R_P$ and so it is clear that $\lambda(P)R_P$ is a two-sided ideal of R_P.

Next we must show that $\lambda^{-1}(\lambda(P)R_P) = P$. Clearly $P \subseteq \lambda^{-1}(\lambda(P)R_P)$, so we assume that $u \in \lambda^{-1}(\lambda(P)R_P)$. We have $\lambda(u) = \lambda(p)\lambda(s)^{-1}$, where $p \in P$, $s \in S$. Therefore $\lambda(us) = \lambda(p)$. Let Q be the ring of quotients of R/P; we have, by the universal property of S inverting maps, a map $R_P \to Q$ making commutative the diagram

$$
\begin{array}{ccc}
R & \longrightarrow & R_P \\
\downarrow & & \downarrow \\
R/P & \longrightarrow & Q
\end{array}
$$

so that if $\lambda(us) = \lambda(p)$ we must have in $\bar{R} = R/P$ that $\bar{u}\bar{s} = \bar{p} = 0$. Now $\bar{s}$ is regular in R/P and so it necessarily follows that $\bar{u} = 0$, i.e., that $u \in P$ and this is what we wanted to show.

Finally let I be a two-sided ideal of R_P. Then we know that $I = \lambda(\lambda^{-1}(I))R_P$ (1.4). Assume that $I \not\subseteq \lambda(P)R_P$. Then by 1.4 we have that $\lambda^{-1}(I) \not\subseteq \lambda^{-1}((P)R_P) = P$. But since R is a PI ring and $\lambda^{-1}(I)$ is a two-sided ideal, it follows that $S \cap \lambda^{-1}(I) \neq 0$ (Chapter II, 5.6) and so clearly $I = R_P$.

§2   Trace Formulas

Let R be a simple algebra of degree n. If $a \in R$ we can consider its characteristic polynomial $f(x) = x^n - \mathrm{Tr}(a)x^{n-1} + \sum_{i=0}^{n-2} \alpha_i x^i$; a satisfies $f(x)$. Consider the standard polynomial $S_{n-1}(x_1, \ldots, x_{n-1})$ in $n-1$ variables. We make the following substitution:

$$
x_1 \to 0 = [f(a), b], \qquad x_i \to [a^{n-i}, b], \qquad i \geq 2.
$$

We obtain

$$
0 = S_{n-1}([a^n, b] - \mathrm{Tr}(a)[a^{n-1}, b] + \sum_{i=0}^{n-2} \alpha_i[a^i, b], [a^{n-2}, b], \ldots)
$$

$$
= S_{n-1}([a^n, b], \ldots) - \mathrm{Tr}(a)S_{n-1}([a^{n-1}, b], [a^{n-2}, b], \ldots)
$$

$$
+ \sum_{i=0}^{n-2} \alpha_i S_{n-1}([a^i, b], [a^{n-2}, b], \ldots).
$$

Now if $i \leqslant n - 2$ we have $S_{n-1}([a^i, b], [a^{n-2}, b], \ldots) = 0$ because two of the arguments in the standard polynomial coincide. We deduce the following identity:

$$S_{n-1}([a^n, b], [a^{n-2}, b], \ldots) = \text{Tr}(a)S_{n-1}([a^{n-1}, b], [a^{n-2}, b], \ldots)$$

which we will write, for simplicity, $\varphi(a, b) = \text{Tr}(a)\psi(a, b)$.

Lemma 2.1   We have $\varphi(a, b) = S_{n-1}([a^n, b], [a^{n-2}, b], \ldots) \neq 0$ for suitable elements $a, b \in R$, provided the center of R has more than n elements.

Proof   If R is infinite with center Z then $S_{n-1}([x^n, y], [x^{n-2}, y], \ldots)$ is an identity in R if and only if it is an identity for $n \times n$ matrices over any field; on the other hand, if R is finite then $R \simeq (Z)_n$, where Z is a finite field. In conclusion it is enough to show that $S_{n-1}([x^n, y], [x^{n-2}, y], \ldots)$ is not an identity for $(Z)_n$ as soon as Z has more than n elements.

So in this case choose $n - 1$ distinct elements $\xi_1, \ldots, \xi_{n-1}$ and a further element ξ_n different from the others, such that $\sum_{i=1}^n \xi_i \neq 0$ (this is clearly possible). Let

$$a = \begin{pmatrix} \xi_1 & & & 0 \\ & \xi_2 & & 0 \\ & & \ddots & \\ 0 & & & \xi_n \end{pmatrix}, \qquad b = \begin{pmatrix} 0 & 1 & 0 & \cdots & 0 \\ 0 & 0 & 1 & \cdots & 0 \\ \vdots & & & \ddots & \\ 0 & & & & 1 \\ 1 & 0 & \cdots & & 0 \end{pmatrix};$$

we have $\text{Tr}(a) = \sum_{i=1}^n \xi_i \neq 0$ so it is enough to prove that $S_{n-1}([a^{n-1}, b], [a^{n-2}, b], \ldots) \neq 0$ to have also $S_{n-1}([a^n, b], [a^{n-2}, b], \ldots) \neq 0$. Now

$$[a^k, b] = \begin{pmatrix} 0 & \xi_1^k - \xi_2^k & \cdots & & 0 \\ & 0 & \xi_2^k - \xi_3^k & & \\ & & & \ddots & \\ 0 & & \cdots & & \xi_{n-1}^k - \xi_n^k \\ \xi_n^k - \xi_1^k & 0 & \cdots & & 0 \end{pmatrix}.$$

Assume we have $n - 1$ elements:

$$u_i = \begin{pmatrix} 0 & \rho_{1i} & 0 & \cdots & 0 \\ 0 & & \rho_{2i} & \cdots & 0 \\ 0 & & & & \vdots \\ 0 & & \cdot & \cdot & \rho_{n-1\ i} \\ \rho_{n\,i} & & \cdot & \cdot & 0 \end{pmatrix}$$

with $i = 1, \ldots, n - 1$; then it is easy to see that the matrix $u_1 u_2 \ldots u_{n-1}$ has in the $1n$ place the entry $\rho_{11}\rho_{22} \cdots \rho_{n-1\ n-1}$. Therefore $S_{n-1}([a^{n-1}, b]$, $[a^{n-2}, b], \ldots)$ has in the $1n$ place the entry

$$\Delta = \det \begin{vmatrix} \rho_{11} & \cdots & \rho_{1\ n-1} \\ & \vdots & \\ \rho_{n-1\ 1} & \cdots & \rho_{n-1\ n-1} \end{vmatrix}$$

where $\rho_{ik} = \xi_i^{\,k} - \xi_{i-1}^k$. Hence $\Delta = \prod_{i<j} (\xi_i - \xi_j) \neq 0$ and so the claim is proved.

Now, as before, let R be a simple algebra of degree n, and Z its center. We consider the reduced norm map $N: R \to Z$; $N(r) \neq 0$ if and only if r is invertible in R. If $u_1, \ldots, u_{n^2}$ is a basis of R over Z we have $N(\sum \alpha_i u_i) = f(\alpha_1, \ldots, \alpha_{n^2})$ where f is a homogeneous polynomial of degree n. It follows that if $a = \sum \alpha_i u_i$, $b = \sum \beta_j u_j$ then $N(\varphi(a, b)) = g(\alpha_i, \beta_j)$ where g is a polynomial of a certain fixed degree h (depending only on n).

Since $\varphi(x, y)$ is not a polynomial identity for simple algebras of degree n it follows (Chapter III, **1.6**) that $N(\varphi(a, b)) = g(\alpha_i, \beta_j)$ is a polynomial formally nonzero, so actually nonzero as soon as the center Z of R has more than h elements.

§3   Characterization of Azumaya Algebras by Identities

The theorem that we want to discuss is a powerful generalization of Chapter II, **1.1** and it describes very precisely the structure of a certain class of *PI* rings, including primitive algebras.

Definition 3.1   A ring R will be called an A_n ring if

(1) R satisfies the identities of $n \times n$ matrices.
(2)   No (nonzero) homomorphic image of R satisfies the identities of $n - 1 \times n - 1$ matrices.

Remark   Under hypothesis (1) the second condition means exactly that $\operatorname{Spec}(R) = \operatorname{Spec}_n(R)$. In fact if R/I satisfies the identities of $n - 1 \times n - 1$ matrices then taking a prime ideal $P \supset I$ we have $P \in \operatorname{Spec}_k(R)$ for some $k \leqslant n - 1$. Conversely if R satisfies (1) then

$$\operatorname{Spec}(R) = \bigcup_{k \leqslant n} \operatorname{Spec}_k(R)$$

and if $\operatorname{Spec}_k(R) \neq 0$ for some $k < n$ we have, taking a $P \in \operatorname{Spec}_k(R)$, that R/P satisfies the identities of $n - 1 \times n - 1$ matrices.

Proposition 3.2   If $\varphi \colon R \to S$ is a central extension and R is an A_n ring then S is also an A_n ring.

Proof   We first have to show that S satisfies all the polynomial identities of $n \times n$ matrices. Now R satisfies all such identities, so R is a quotient of a suitable large free algebra $\mathbf{Z}\{x_s\}$ modulo the ideal I_n of all the identities of $n \times n$ matrices. The ring $\mathbf{Z}\{x_s\}/I_n$ is isomorphic to a subring of a matrix ring $(\mathbf{Z}[\xi_{ijs}])_n$. Now let W be the center of S and let $\mathbf{Z}[u_t]$ be a large polynomial ring having W as its homomorphic image. Then $S = \varphi(R)W$ is a homomorphic image of $\mathbf{Z}\{x_s\}/I_n \otimes_{\mathbf{Z}} \mathbf{Z}[u_t]$ which is a subring of $(\mathbf{Z}[\xi_{ijs}] \otimes_{\mathbf{Z}} \mathbf{Z}[u_t])_n$. This last ring is a ring of $n \times n$ matrices over a commutative ring, so finally S satisfies all the identities of this ring and so of $n \times n$ matrices.

Next we must verify that no homomorphic image of S satisfies the identities of $n - 1 \times n - 1$ matrices. Let, therefore, $\bar{S}$ be such a homomorphic image. The image $\bar{R}$ of R in $\bar{S}$ then would satisfy also all the identities of $n - 1 \times n - 1$ matrices, contradicting the assumption that R is an A_n ring.

The theorem we have in mind is the following.

Theorem 3.3 (M. Artin)   The two following statements for a ring R are equivalent:

(1) R is an A_n ring.
(2) R is an Azumaya algebra of (constant) rank n^2 over its center.

Proof   The proof will be split in several parts.

First we prove that $(2) \Rightarrow (1)$. We have to show that an Azumaya algebra R of rank n^2 over its center A satisfies (1) and (2) of Definition **3.1.** For (1) we know that there exists a faithfully flat extension $A \to B$ of the center A of R such that $R \otimes_A B \simeq (B)_n$. Therefore $R \subseteq (B)_n$ satisfies the identities of $n \times n$ matrices. For (2) let P be a prime ideal of R. By the structure theory of Azumaya algebras we know that there exists a prime ideal $\mathbf{p}$ in A such that $P = \mathbf{p}A$. Consider $R/P \simeq R \otimes A/\mathbf{p}$ and let F be the field of quotients of $A/\mathbf{p}$, then $R/P \subseteq R \otimes_A F$ and $R \otimes_A F$ is an Azumaya algebra of rank n^2 over F and so a central simple algebra of degree n. R/P is an order in $R \otimes_A F$ and it follows that $P \in \mathrm{Spec}_n(R)$ so that $\mathrm{Spec}(R) = \mathrm{Spec}_n(R)$.

We now come to the main part of the theorem, i.e., $(1) \Rightarrow (2)$. We will work out the validity of the theorem first in several special cases.

(a) R is a projective module of rank n^2 over its center A. In this case, to prove that R is an Azumaya algebra we only have to prove that for every maximal ideal $\mathbf{m} \subset A$ the ring $R/\mathbf{m}R$ is a central simple algebra over $A/\mathbf{m}$. Assume by contradiction that $R/\mathbf{m}R$ were not central simple over $A/\mathbf{m}$. Take a simple homomorphic image S of $R/\mathbf{m}R$ and its center Z; we will clearly have $\dim_Z < n^2$. Thus it follows that S satisfies the identities of $n - 1 \times n - 1$ matrices contradicting the hypothesis that R should be an A_n ring.

(b) R is a prime local algebra with maximal ideal M over a ring Λ, and the ring $\overline{\Lambda} = \Lambda/\Lambda \cap M$ has more than $h + 1$ elements (h is the number considered in §2). In this case, since R is an A_n ring, we have $(0), M \in \mathrm{Spec}_n(R)$. Let Q be the ring of quotients of R and Z its center. Q is a central simple algebra of rank n^2 over Z. Let T be the set of regular elements modulo M. Since R is an A_n ring we can satisfy the hypothesis of **1.6**(1) choosing the multilinear identity

$$S_{2n}(x_1, \ldots, x_{2n}) = x_1 S_{2n-1}(x_2, \ldots, x_{2n}) + \varphi(x_1, \ldots, x_{2n}).$$

Therefore we see by **1.6** that if $r \in T$ the ideal $I = \{t \in R | Rt \subseteq rR\}$ is a two-sided ideal and $I \not\subset M$. Therefore, as R is local, we must have $I = R$ and so r is invertible in R and $R = R_M$.

If $a, b \in R$ we have $\varphi(a, b) = \text{Tr}(a)\psi(a, b)$ (§2). Further, if $c \in R$ and $\lambda \in \Lambda$ we have

$$\varphi(\lambda a + c, b) = [\lambda\text{Tr}(a) + \text{Tr}(c)]\psi(\lambda a + c, b).$$

Choose now a particular $a \in R$ and consider the algebra R/M. Since $\overline{\Lambda} = \Lambda/\Lambda \cap M$ has more than $h + 1 > n$ elements we know that the center $\overline{Z}$ of R/M has more than h elements. Therefore we can find $c, b \in R$ such that $0 \neq N(\psi(\bar{c},\bar{b})) \in \overline{Z}$ (§2). Now $N(\psi(t\bar{a} + \bar{c},\bar{b})) = f(t)$ is a polynomial in t of degree less or equal to h and $f(0) = N(\psi(\bar{c}, \bar{b})) \neq 0$ so $f(t)$ is a non-zero polynomial. Since $\overline{\Lambda}$ has more than $h + 1$ elements we can find $\lambda \in \Lambda$ such that $0 \neq \lambda \in \Lambda$ and $N(\psi(\bar{\lambda}\bar{a} + \bar{c}, \bar{b})) \neq 0$. Therefore λ, $\psi(\lambda a + c, b)$ and $\psi(c, b)$ are regular elements modulo M and so, by the previous remarks, we have λ^{-1}, $\psi(\lambda a + c, b)^{-1}$, $\psi(c, b)^{-1} \in R$, and therefore

$$\lambda\text{Tr}(a) + \text{Tr}(c) = \varphi(a + \lambda c, b)\psi(a + \lambda c, b)^{-1} \in R,$$

$$\text{Tr}(c) = \varphi(c, b)\psi(c, b)^{-1} \in R,$$

and finally

$$\text{Tr}(a) = \lambda^{-1}\{[\lambda\text{Tr}(a) + \text{Tr}(c)] - \text{Tr}(c)\} \in R.$$

Further, passing to $\overline{R} = R/M$ we see that

$$\text{Tr}(\bar{a}) = \bar{\lambda}^{-1}\{[\bar{\lambda}\text{Tr}(\bar{a}) + \text{Tr}(\bar{c})] - \text{Tr}(\bar{c})\}$$
$$= \bar{\lambda}^{-1}\{\varphi(\bar{\lambda}\bar{a} + \bar{c}, \bar{b})\psi(\bar{\lambda}\bar{a} + \bar{c}, \bar{b})^{-1} - \varphi(\bar{c}, \bar{b})\psi(\bar{c}, \bar{b})^{-1}\} = \overline{\text{Tr}(a)},$$

so we have proved that the map $\text{Tr}: R \to Q$ is in fact a map $\text{Tr}: R \to R$ and the following diagram is commutative:

$$
\begin{array}{ccc}
R & \xrightarrow{\text{Tr}} & R \\
\downarrow & & \downarrow \\
R/M & \xrightarrow{\text{Tr}} & R/M
\end{array}
$$

Now let $u_1, \ldots, u_{n^2} \in R$ be such that the $\bar{u}_i$'s are a basis of R/M over $\bar{Z}$. This means exactly that $d = \det(\mathrm{Tr}(\bar{u}_i \bar{u}_j)) \neq 0$; but if this is true we see that $\Delta = \det(\mathrm{Tr}(u_i u_j))$ maps into d modulo M and so is regular modulo M and therefore $\Delta^{-1} \in R$. Then it is clear by the usual argument that R is free of rank n^2 over its center with the u_i's its basis. Then we can finally apply case (a) and prove that R is a rank n^2 Azumaya algebra over its center.

(c)   Let R be a prime A_n ring and assume that for all maximal ideals M of R the ring R_M is an Azumaya algebra over its center; then R is a rank n^2 Azumaya algebra. Since R is an A_n ring M, $(0) \in \mathrm{Spec}_n(R)$ and so, by the same remarks as in case (b), we can localize R at M. Further, if we show that the elements of R regular modulo M are regular in R we will have by $\mathbf{1.5}$(2) that R_M is contained in the total ring of quotients of R. The fact that the elements of R regular modulo M are regular in R depends on the following remark. Let R be prime of degree n, with Q its ring of quotients and Z the center of Q. Let, further, P be a prime ideal of R; $R/P = \bar{R}$, $\bar{Q}$ the ring of quotients of $\bar{R}$ and $\bar{Z}$ the center of $\bar{Q}$. Assume that $\bar{Q}$ is also of degree n. If $c \in R$ is regular modulo P then $\bar{c}\bar{R}\bar{Z} = \bar{c}\bar{Q} = \bar{Q}$ so $\bar{c}\bar{R}$ satisfies the same multilinear identity as $\bar{Q}$. Now if c were not regular in R we would have $cRZ \neq Q$ and so cRZ would satisfy a multilinear identity that is not satisfied by any simple algebra of degree n (Chapter I, $\mathbf{5.3}$). This is impossible because it would imply that $\bar{c}\bar{R}$ and so also $\bar{Q}$ should satisfy this identity. Therefore we now have, in the case under consideration, that $R_M \subseteq Q$ for all maximal ideals M of R. Let Z_M be the center of R_M. Since $RZ = Q$ we clearly have $Z_M \subset Z$ (Chapter II, $\mathbf{6.7}$). Let A be the center of R; we claim that $A = \{\bigcap Z_M | M$ maximal ideal of $R\}$. Clearly $A \subseteq \bigcap Z_M$. Conversely let $\alpha \in \bigcap Z_M$; we must prove $\alpha \in R$. Let $I = \{r \in R | r\alpha \in R\}$; I is a two-sided ideal of R. Assume that I is proper; then I can be enlarged to a maximal ideal N. Then we have $\alpha \in Z_N$ so that $\alpha = cd^{-1}$, with $c, d \in R$, and d regular modulo N. Now $d\alpha = c \in R$ and so $d \in I \subset N$, a contradiction. Therefore $I = R$ and so $\alpha \in R$, and we have the desired equality.

Consider now $r \in R$, $r \in R_M$ for every maximal ideal M. Therefore $\mathrm{Tr}(r) \in Z_M$, since R_M is an Azumaya algebra over Z_M for every M, and so $\mathrm{Tr}(r) \in \bigcap Z_M = A$. It follows that $\mathrm{Tr}: R \to R$. To show that R is an Azumaya algebra over A it is necessary to show that $RA_{\mathbf{m}}$ is an Azumaya algebra for all maximal ideals $\mathbf{m}$ of A. For this purpose consider the ideal J of R generated by all discriminants of all bases $z_1, \ldots, z_{n^2}$ of Q over Z contained in the ring R. We claim first of all that $J = R$. Assume by contradiction that $J \neq R$. Then we can find a maximal ideal $M \supset J$. Consider the Azumaya algebra R_M which is free over its center Z_M [since R_M is local, Z_M is local and so R_M is Z_M-free, or also by case (b)]. Let $u_1, \ldots, u_{n^2} \in R_M$ be a basis over Z_M; $u_1, \ldots, u_{n^2}$ is also a basis for Q over Z. We can find an invertible element s of R_M such that $u_i s \in R$, since every element of R_M is of the form at^{-1}, with $a, t \in R$. The elements $\bar{u}_1 = u_1 s, \ldots, \bar{u}_{n^2} = u_{n^2} s$ are again a basis of R_M over Z_M.

The discriminant of such a basis is invertible in Z_M and so it cannot be contained in $J \subseteq M$; this contradicts the hypothesis $J \neq R$. Therefore we have now that $J = R$; in particular $1 = \sum r_i d_i$ where $r_i \in R$ and d_i is a discriminant of a basis of Q over Z contained in R.

Consider for $a \in Q$ the regular representation $a_R \colon Q \to Q$, with $a_R(u) = ua$. Det $(a_R) \in Z$ and if $a = \sum x_i r_i$, with $x_i \in A$, $r_i \in R$, then $\det(a_R)$ is a homogeneous polynomial in the x_i's with coefficients in Z. Since $r_i \in R$, restricting the regular representation to each R_M which is an Azumaya algebra free over Z_M, we see that these coefficients are in Z_M [25, 40]. Since this is true for every maximal ideal M we see that these coefficients are in A. Let us write $\det(x_i r_i) = \sum x_i g_i(x)$. We have

$$1 = \det(1_R) = \det((\sum d_i r_i)_R) = \sum d_i g_i(d_i) \in \sum d_i A.$$

We have thus proved that the discriminants d_i generate in the ring A the ideal A itself.

Now we can finish from this last remark. We know that A is an order in Z (Chapter II, **5.8**) so $Q \simeq R \otimes_A Z$. Consider the ring $A[1/d_i]$; we have $A[1/d_i] \subseteq Z$ and $R_i = R \otimes_A A[1/d_i] \subseteq Q$. Now R_i is free over $A[1/d_i]$ because the elements of R having d_i as discriminant are a basis of R_i over $A[1/d_i]$. Next R_i is a central extension of R and so by **3.2** it is an A_n ring; then finally by (a) we have that R_i is a rank n^2 Azumaya algebra over $A[1/d_i]$. To deduce from this that R is a rank n^2 Azumaya algebra over A we use the fact that the ideal generated by the d_i's in A is A itself [25].

(d) We are now ready to prove the general case. Let R be an A_n ring; since R satisfies the identities of $n \times n$ matrices, then R is a homomorphic image of $\mathbf{Z}\{x_s\}/I_n = \mathbf{Z}\{\xi_s\} = B$ (where ξ_s represent generic matrices). Let $\lambda \colon B \to R$ be the epimorphism which we have constructed.

Let $L = \operatorname{Ker} \lambda$ and $S = \{u \in B | \lambda(u) = 1\}$. We can apply **1.6(2)** relatively to the identity

$$S_{2n}(x_1, \ldots, x_{2n}) = x_1 S_{2n-1}(x_2, \ldots, x_{2n}) + \varphi(x_1, \ldots, x_{2n}),$$

since, as R is an A_n ring, no homomorphic image of R satisfies

$S_{2n-1}(x_2, \ldots, x_{2n})$. We therefore form the localization B_S of B at S. Since B is an order in a division ring D of rank n^2 over its center, we have $B_S \subseteq D$.

Consider $I_S = IB_S$; we claim that

(i) I_S is a two-sided ideal of B_S.
(ii) $I_S \cap B = I$.
(iii) $B_S/I_S \simeq B/I \simeq R$.

(i) The ideal I_S is clearly a right ideal. Let $as^{-1} \in I_S$, $a \in I$, $s \in S$, and $bt^{-1} \in B_S$. Compute $bt^{-1}as^{-1}$; we observe that $t^{-1}a = cd^{-1}$ for some $c \in B$, $d \in S$ so that $ad = tc$. Computing modulo I we have $\bar{a} = 0$, $\bar{t} = 1$ and so $\bar{c} = 0$, i.e., $c \in I$; therefore $bt^{-1}as^{-1} = cb(sd)^{-1} \in I_S$.

(ii) Clearly $I_S \cap B \supseteq I$; if $as^{-1} = b \in B$ and $a \in I$ we have $a = bs$ and, computing modulo I, we have $0 = \bar{a} = \bar{b}$ so also $b \in I$.

(iii) From what we have seen $B/I \subseteq B_S/I_S$, but clearly $as^{-1} \equiv a$ modulo I_S so $B_S/I_S \simeq B/I \simeq R$.

We need now the following basic remark: B_S is an A_n ring. In fact clearly B_S satisfies all the identities of $n \times n$ matrices (since $B_S \subseteq D$ and D is a simple algebra of degree n). Let now $u(x_1, \ldots, x_k)$ be a polynomial identity for $n - 1 \times n - 1$ matrices but not for $n \times n$ matrices. Consider in R the ideal T generated by the elements $u(r_1, \ldots, r_k)$. R/T satisfies $u(x_1, \ldots, x_k)$; therefore if we had $R/T \neq 0$, every simple homomorphic image of R/T would satisfy $u(x_1, \ldots, x_k)$ and so it would have degree $\leqslant n-1$, contradicting the hypothesis that R is an A_n ring. Therefore $T = R$ and so $1 = \sum_i a_i u(r_1^{(i)}, \ldots, r_k^{(i)}) b_i$ for some suitable $a_i, b_i, r_j^{(i)} \in R$.

Picking elements $\alpha_i, \beta_i, \rho_j^{(i)} \in B$ mapping respectively to $a_i, b_i, r_j^{(i)}$, we see that the element $s = \sum \alpha_i u(\rho_1^{(i)}, \ldots, \rho_k^{(i)}) \beta_i \in S$. In B_S the element s is invertible. If $\bar{B}_S$ had a homomorphic image $\bar{B}_S$ satisfying the identities of $n - 1 \times n - 1$ matrices we would have that $u(x_1, \ldots, x_k)$ vanishes identically on $\bar{B}_S$. This is impossible because it would imply that s is zero in B_S while s is invertible. Therefore B_S is an A_n ring. At this point, if we can prove that B_S is an Azumaya algebra of rank n^2 over its center the same will follow for R, since R is a homomorphic image of B_S.

To prove that B_S is an Azumaya algebra we would like to use all previous cases. Unfortunately we do not know that B_S is an algebra over a commutative ring Λ to allow us to use case (b). We can reduce to that case as follows. Let Λ be the ring of algebraic integers in a cyclotomic field $K = Q(\zeta)$ with ζ a primitive nth root of 1 and Q the field of rationals. If we choose a suitable n we can ensure that the residue fields of Λ all have more than h elements. In fact consider all primes $p_1, \ldots, p_k$ such that $p_i \leqslant h + 1$ and let n be prime to $p_1, \ldots, p_k$ and such that $p_i^{h+1} < n$; then the residue fields of Λ of characteristic $p > h + 1$ have surely more than $h + 1$ elements while if the characteristic is $p_i \leqslant h + 1$ then p_i is unramified in Λ and the residue field has p_i^{j} elements, where j is the minimal number such that $p_i^{j} \equiv 1 \pmod{n}$. In particular $h + 1 < p_i^{h+1} < n < p_i^{j}$.

We consider the ring $C = B_S \otimes_Z \Lambda$; C is an A_n ring by **3.2** and further $C \subset D \otimes_Z \Lambda \subset D \otimes_Q (\Lambda \otimes_Z Q) = D \otimes_Q Q(\zeta)$. But $D \otimes_Q Q(\zeta)$ is the division ring generated by the generic matrices over $Q(\zeta)$ and therefore C is a domain (in particular a prime ring). Now if M is a maximal ideal of C the ring C_M is a local prime A_n ring and an algebra over Λ. For all prime ideals P of Λ we have that Λ/P has more than $h + 1$ elements. C_M is Azumaya by case (b), therefore by case (c) C is also a rank n^2 Azumaya algebra over its center. Now Λ is free over Z and so this implies that B_S is also a rank n^2 Azumaya over its center. This terminates the proof of the theorem.

Chapter VIII

THE CENTER OF A *PI* RING

§1   The Center of a *PI* Ring

If R is a *PI* ring, what can we say about the center of R? In general not very much, as simple examples taking triangular matrices show.

We ask how to find polynomials which evaluated in $n \times n$ matrices yield elements of the center.

Definition 1.1   Given a commutative ring Λ, a central polynomial for $n \times n$ matrices with coefficients in Λ is a polynomial $G(x_1, \ldots, x_m) \in \Lambda\{x_1, \ldots, x_m\}$ which, whenever evaluated in $(A)_n$ (with A a commutative Λ algebra), gives elements in the center.

Remark   Any polynomial identity of $n \times n$ matrices is a central polynomial. We seek nontrivial central polynomials, i.e., central polynomials which are not constant.

Proposition 1.2

(1) $G(x_1, \ldots, x_m)$ is a central polynomial for $n \times n$ matrices if and only if, evaluated in the generic matrices $\xi_1, \ldots, \xi_m$, it lies in the center of $\Lambda\{\xi_1, \ldots, \xi_m\}$, $(m \geqslant 2)$.

(2)   If $G(x_1, \ldots, x_m)$ is a central polynomial for $n \times n$ matrices then $G(H_1, \ldots, H_m)$ is again a central polynomial for all choices of polynomials $H_1, \ldots, H_m$.

(3)   The central polynomials for $n \times n$ matrices are a subring of $\Lambda\{x_1, \ldots, x_m\}$.

(4)   If $G(x_1, \ldots, x_m)$ is a central polynomial with no constant term then it is a polynomial identity for $n - 1 \times n - 1$ matrices.

Proof   For (2) and (3) the proofs are trivial. For (1) the proof is quite easy since, if G is a central polynomial for $n \times n$ matrices, it will yield values in the center whenever computed in any subalgebra of an algebra of $n \times n$ matrices over a commutative Λ algebra. In particular $G(\xi_1, \ldots, \xi_m)$ is central in $\Lambda\{\xi_1, \ldots, \xi_m\}$. Conversely if $G(\xi_1, \ldots, \xi_m)$ is in the center of $\Lambda\{\xi_1, \ldots, \xi_m\}$ then adding an extra generic matrix ξ_{m+1} we have

$$\Lambda\{\xi_1, \ldots, \xi_m\} \subseteq \Lambda\{\xi_1, \ldots, \xi_{m+1}\}.$$

This is not an extension, but passing to the quotient division ring it becomes a central extension

$$\Lambda\langle\xi_1, \ldots, \xi_m\rangle \subseteq \Lambda\langle\xi_1, \ldots, \xi_{m+1}\rangle$$

(since $m \geqslant 2$). In particular any element in the center of $\Lambda\{\xi_1, \ldots, \xi_m\}$ is also in the center of $\Lambda\{\xi_1, \ldots, \xi_{m+1}\}$. Therefore

$$[G(\xi_1, \ldots, \xi_m), \xi_{m+1}] = 0.$$

This means that $[G(x_1, \ldots, x_m), x_{m+1}]$ is a polynomial identity for $n \times n$ matrices. But this is precisely the condition that $G(x_1, \ldots, x_m)$ is a central polynomial.

(4)   Let $G(x_1, \ldots, x_m)$ be a polynomial with no constant term. If we compute it in $(A)_{n-1}$ and consider $(A)_{n-1} \subseteq (A)_n$ we see that we always obtain scalars of $(A)_n$. But the only scalar of $(A)_n$ contained in $(A)_{n-1}$ is zero and (4) is proved.

As a corollary we see that in the mapping

$$\varphi \colon \Lambda\{x_1, \ldots, x_m\} \to \Lambda\{\xi_1, \ldots, \xi_m\},$$

the subring of central polynomials for $n \times n$ matrices is exactly the pre-image of the center of $\Lambda\{\xi_1, \ldots, \xi_m\}$.

The concept just developed would be void of interest if we could not show the existence of nontrivial central polynomials.

Fortunately such polynomials exist in a certain abundance.

The first important result will be given therefore by the following construction, owing to Formanek [35].

Let n be a natural number and consider the following polynomial over **Z** in the commutative variables $x_1, \ldots, x_{n+1}$:

$$f(x_1, \ldots, x_{n+1}) = \prod_{i=2}^{n} (x_1 - x_i)(x_{n+1} - x_i) \prod_{\substack{i=2 \\ i<j}}^{n} (x_i - x_j)^2.$$

This polynomial computed for $x_{n+1} = x_1$ gives $\prod_{i=1, j<i}^{n} (x_i - x_j)^2 = D(x_1, \ldots, x_n)$, the discriminant of $x_1, \ldots, x_n$. We expand f and we get $f(x_1, \ldots, x_{n+1}) = \sum a_{r_1 \ldots r_{n+1}} x_1^{r_1} x_2^{r_2} \ldots x_{n+1}^{r_{n+1}}$.

We consider now the noncommutative polynomial in the variables $X, Y_1, Y_2, \ldots, Y_n$:

$$F(X, Y_1, Y_2, \ldots, Y_n) = \sum a_{r_1 r_2 \ldots r_{n+1}} X^{r_1} Y_1 X^{r_2} Y_2 \ldots Y_n X^{r_{n+1}}.$$

F is linear in the Y_i's. Let us compute F when X is a diagonal matrix $X = \text{diag}(x_1, \ldots, x_n)$, and $Y_j = e_{i_j t_j}$ is a matrix unit; we have

$$\bar{X}^{r_1} e_{i_1 t_1} \bar{X}^{r_2} e_{i_2 t_2} \ldots e_{i_n t_n} \bar{X}^{r_{n+1}} = 0 \qquad \text{unless} \quad t_1 = i_2, t_2 = i_3, \text{ etc.}$$

If we compute with $\bar{Y}_j = e_{i_j i_{j+1}}$ we get

$$F(\bar{X}, e_{i_1 i_2}, e_{i_2 i_3}, \ldots, e_{i_n i_{n+1}}) = f(x_{i_1}, x_{i_2}, \ldots, x_{i_{n+1}}) e_{i_1 i_{n+1}}.$$

Now f vanishes if two of the entries are equal, except if the first is equal to the last. Therefore $F(\bar{X}, e_{i_1 i_2}, \ldots, e_{i_n i_{n+1}})$ vanishes unless $i_1, \ldots, i_n$ are a permutation $\sigma(1), \sigma(2), \ldots, \sigma(n)$ of $1, 2, \ldots, n$ and $i_{n+1} = i_1$. In this case,

$$F(X, e_{\sigma(1)\sigma(2)}, e_{\sigma(2)\sigma(3)}, \ldots, e_{\sigma(n)\sigma(1)}) = f(x_{\sigma(1)}, x_{\sigma(2)}, \ldots, x_{\sigma(n)}, x_{\sigma(1)})$$

$$= D(x_{\sigma(1)}, \ldots, x_{\sigma(n)}) e_{\sigma(1)\sigma(1)} = D(x_1, \ldots, x_n) e_{\sigma(1)\sigma(1)},$$

since $D(x)$ is a symmetric polynomial.

Now consider

$$G(X, Y_1, \ldots, Y_n) = F(X, Y_1, \ldots, Y_n) + F(X, Y_2, \ldots, Y_n, Y_1)$$
$$+ F(X, Y_3, \ldots, Y_1, Y_2) + \cdots + F(X, Y_n, Y_1, \ldots, Y_{n-1}).$$

When we compute it with $\mathrm{diag}(x_1, \ldots, x_n)$ and matrix units $Y_j = e_{i_j t_j}$, from the previous argument we get zero, unless the units are as before; in this case we get

$$G(X, e_{\sigma(1)\sigma(2)}, \ldots, e_{\sigma(n)\sigma(1)}) = \sum_{i=1}^{n} D(x_1, \ldots, x_n) e_{\sigma(i)\sigma(i)}$$

$$= D(x_1, \ldots, x_n) 1_n,$$

where 1_n is the identity matrix of $n \times n$ matrices.

More generally if $Y_i = (y_{st}^{(i)})$ we have, by the multilinearity of G in the Y_i's,

$$G(X, Y_1, \ldots, Y_n) = D(x_1, \ldots, x_n) \psi(y_{st}^{(i)}) 1_n,$$

where $\psi(y_{st}^{(i)})$ is a certain polynomial that could be written explicitly.

Theorem 1.3 (Formanek)   The polynomial $G(X, Y_1, \ldots, Y_n)$ gives a nontrivial central polynomial for $n \times n$ matrices (in any characteristic).

Proof   By **1.2** it is enough to show that $G(\xi, \xi_1, \ldots, \xi_n)$ is in the center of $\Lambda\{\xi, \xi_1, \ldots, \xi_n\}$, with the ξ_i generic matrices. Now since ξ is generic it can be brought to diagonal form, conjugating with some matrix A. Conjugating we have $\xi', \xi_1', \ldots, \xi_n'$, with $\xi' = \mathrm{diag}(x_1, \ldots, x_n)$. Therefore

$$G(\xi', \xi_1', \ldots, \xi_n') = D(x_1, \ldots, x_n) \psi(\xi_{ijt}') 1_n.$$

Now

$$G(\xi, \xi_1, \ldots, \xi_n) = A^{-1} G(\xi', \xi_1', \ldots, \xi_n') A = D(x_1, \ldots, x_n) \psi(\xi_{ij,t}') 1_n.$$

The fact that G is now nontrivial has already been shown since, if we compute G in $(A)_n$ and A is a field with more than n elements, then it is enough to take X diagonal with distinct entries and the Y_i's suitable matrix units to get a nonzero value.

We want to deduce some interesting consequences of the previous theorem.

Theorem 1.4   Let R be a prime *PI* ring, Q its ring of fractions, Z the center of Q, and A the center of R. Then Z is the field of fractions of A and $Q \simeq R \otimes_A Z$.

Proof   The ring R will have some degree n and we consider the polynomial $G(X, Y_1, \ldots, Y_n)$ of the previous theorem. We can assume that R is infinite, otherwise $R = Q$. Since R is infinite and $G(X, Y_1, \ldots, Y_n)$ is not an identity for $n \times n$ matrices, G does not vanish on R. Therefore it assumes some nonzero value α in the center of R. Consider

$$R\left[\frac{1}{\alpha}\right] = \left\{\frac{r}{\alpha^k}\middle| r \in R,\ k \in N\right\}.$$

Then $R[1/\alpha]$ is a ring satisfying all the identities of $n \times n$ matrices; α is the value in R of a polynomial identity for $n - 1 \times n - 1$ matrices (**1.2**) and since α is invertible in $R[1/\alpha]$ we see that no homomorphic image of $R[1/\alpha]$ satisfies the identities of $n - 1 \times n - 1$ matrices. We can apply, therefore, Chapter VII, **3.3** and deduce that $R[1/\alpha]$ is an Azumaya algebra over its center B. Now clearly $B = A[1/\alpha]$ and, since $R[1/\alpha]$ is an Azumaya algebra over B, we have that Z is the total ring of fractions of B, therefore also of A.

The fact that $Q \simeq R \otimes_A Z$ follows by the remark that we have a map $R \otimes_A Z \to Q$ given by $r \otimes z \to rz$. This map is easily seen to be injective and it is surjective since $Q = RZ$ (Chapter II, **5.7**).

Another consequence is the following. For the sake of simplicity let Λ be an infinite field. Consider $R = \Lambda\{\xi_1, \ldots, \xi_m\}$ the ring of generic $n \times n$ matrices, and let Z be the center of R. Z is clearly a graded subring with Λ in degree zero. Let Z^+ be the part in positive degree of Z. By **1.2** we know that Z^+ is made of polynomial identities of $n - 1 \times n - 1$ matrices. Let $I = Z^+ R$ be the ideal generated by Z^+ and N the nil radical of I. We have the following theorem.

Theorem 1.5 N is the image in R of the ideal of polynomial identities of $n - 1 \times n - 1$ matrices.

Proof   Clearly N is contained in the ideal of polynomial identities of $n - 1 \times n - 1$ matrices. If $\tilde{N}$ is the preimage of N in $\Lambda\{x_1, \ldots, x_m\}$ we see that $\tilde{N}$ is a radical T ideal, $\tilde{N}$ is contained in the T ideal of identities of $n - 1 \times n - 1$ matrices and contains properly the T ideal of the identities of $n \times n$ matrices. Therefore by Chapter III, **2.3** $\tilde{N}$ is the ideal of identities of $n - 1 \times n - 1$ matrices.

In particular $\Lambda\{\xi_1, \ldots, \xi_m\}/N$ is isomorphic to the ring of generic $n - 1 \times n - 1$ matrices.

§2   The Formanek Center

If $\mathscr{C}_n$ is the variety of rings satisfying the identities of $n \times n$ matrices we can interpret some of our results in the following categorical form.

If $R \in \mathscr{C}_n$ we define $F(R)$ (the Formanek center of R) to be the subring of R obtained by evaluating all the central polynomials, without constant term, in R. $F(R)$ enjoys the following properties.

Theorem 2.1
  (1) $F(R)$ is a functor in $\mathscr{C}_n$.
  (2) $F(R)$ is contained in the center of R.
  (3)   If $\psi: R \to S$ is surjective then $F(R) \to F(S)$ is surjective.
  (4)   A representation (not necessarily unit-preserving) $\varphi: R \to (K)_n$, with K a field, is an *AI* representation if and only if $\varphi(F(R)) \neq 0$.
  (5) R is a rank n^2 Azumaya algebra over its center if and only if $F(R)R = R$.
  (6) $\sum_{n-1} (R) = V(F(R))$ (in Spec(R)).

Proof   (1) is clear and (2) follows immediately from **1.2**.
  (3)   The proof is quite simple. If $f(x_1, \ldots, x_s)$ is a central polynomial and $a_1, \ldots, a_s \in S$ then $f(a_1, \ldots, a_s)$ is a typical generator of the subring $F(S)$. Since ψ is surjective there are elements $c_1, \ldots, c_s \in R$ such that $\psi(c_i) = a_i$ and then $f(c_1, \ldots, c_s) \in F(R)$ and $\psi(f(c_1, \ldots, c_s)) = f(a_1, \ldots, a_s)$.
  (4)   If φ is an *AI* representation then $\varphi(R)$ is a degree n prime ring, so it is either a finite matrix algebra or it satisfies exactly the identities of $n \times n$ matrices. In either case we know that we can evaluate a suitable

central polynomial (e.g., the one given in **1.3**) and obtain a nonzero value. Therefore $\varphi(F(R)) \neq 0$. Suppose now that $\varphi(R)K \neq (K)_n$. $\varphi(R)K$ is a finite-dimensional algebra and $\dim_K \varphi(R)K < n^2$. If J denotes the Jacobson radical of $\varphi(R)K$ it is clear that $\varphi(R)K/J$ satisfies all the identities of $n - 1 \times n - 1$ matrices. Since the central polynomials without constant term are identities of $n - 1 \times n - 1$ matrices we see that $\varphi(F(R)) \subseteq J$. On the other hand $\varphi(F(R)) \subseteq F((K)_n) = K$. Now J is nilpotent and so $J \cap K = 0$; this shows that $\varphi(F(R)) = 0$.

(5)   Let M be a maximal ideal of R. Consider $\bar{R} = R/M$. If $M \in \operatorname{Spec}_n(R)$ then there is an AI representation $\varphi: R \to (K)_n$ with kernel M; otherwise there is a non-AI representation $\psi: R \to (K)_n$ (not preserving unit). R is a rank n^2 Azumaya algebra if and only if $M \in \operatorname{Spec}_n(R)$ for every maximal ideal M (Chapter VII, **3.3**). Now if $F(R)R \neq R$ there is a maximal ideal $M \supset F(R)R$. Since in the map $R \to R/M$, $F(R)$ is sent to zero, M is not the kernel of an AI representation; therefore $M \notin \operatorname{Spec}_n(R)$.

Conversely if $F(R)R = R$ and M is any maximal ideal then the image of $F(R)$ in R/M is nonzero, so M is the kernel of an AI representation; therefore $M \in \operatorname{Spec}_n(R)$.

(6)   This has been proved during the proof of (5).

Now let R be a Λ algebra and consider $Z(R) = \Lambda + F(R)$ (we add the constants to $F(R)$). We claim the next theorem.

Theorem 2.2

(1)   If $\alpha \in F(R)$ then $R \otimes_{Z(R)} Z(R)[1/\alpha]$ is an Azumaya algebra.

(2)   The open set $\tilde{U}_R$ of $\operatorname{Spec}(Z(R))$ given by $\operatorname{Spec}(Z(R)) - V(F(R))$ is endowed in this way with a sheaf of Azumaya algebras. It is isomorphic to the space U_R of Chapter IV, **1.8** with its canonical sheaf of Azumaya algebras.

Proof   (1)   If $S = R \otimes_{Z(R)} Z(R)[1/\alpha]$ then $\alpha \in F(S)$ so $S = F(S)S$ and (1) follows from the previous theorem.

(2)   If $\varphi: R \to B$ is an AI representation of R in a rank n^2 Azumaya algebra over its center A, then we claim that $\varphi(F(R))A = A$. In fact if m is any maximal ideal of A we have that B/mB is a rank n^2 simple algebra with center A/m and $R \xrightarrow{\varphi} B \xrightarrow{\pi} B/mB$ is an AI representation. Therefore $\pi\varphi(F(R)) \neq 0$ so $\varphi(F(R)) \not\subseteq m$. Therefore, $\varphi(F(R))A = A$. This shows that the induced map $\varphi: Z(R) \to A$ gives rise to a map $\varphi^*: \operatorname{Spec}(A) \to \tilde{U}$.

Conversely, if A is a ring and $\psi^*: \operatorname{Spec}(A) \to U$ is a map of schemes, ψ^* is given by a ring homomorphism $\psi: Z(R) \to A$ with $\psi(F(R))A = A$. This implies that the algebra $R \otimes_{Z(R)} A$ is an Azumaya algebra by (1) and the local nature of Azumaya algebras. Therefore since $\tilde{U}_R$ classifies the same functor as U_R we have $\tilde{U}_R \simeq U_R$ in a canonical way.

As a consequence we see the following.

Corollary 2.3   If $R \in \mathscr{C}_n$ then $\tilde{U}_R$ is homeomorphic to $\operatorname{Spec}_n(R)$. Its sheaf has stalk isomorphic to R_P at P (with $P \in \operatorname{Spec}_n(R)$).

Proof   We have essentially proved that the map $\operatorname{Spec}(R) \to \operatorname{Spec}(Z(R))$ induced by the inclusion $Z(R) \to R$ induces a bijection $j: \operatorname{Spec}_n(R) \to \tilde{U}_R$. This is continuous by construction. To see that it is also a closed map let I be an ideal of R and $V(I)$ the closed set of $\operatorname{Spec}_n(R)$ induced by I. We claim that $j(V(I)) = V(I \cap Z(R)) \cap \tilde{U}_R$. In fact it is clear that $j(V(I)) \subset V(I \cap Z(R)) \cap \tilde{U}_R$. Conversely if $P \in \operatorname{Spec}_n(R)$ and $P \notin V(I)$ then the image $\bar{I}$ of I in R/P is nonzero. Then $\bar{I}$ is a prime ring of degree n and so $F(\bar{I}) \neq 0$. Now $F(I)$ maps onto $F(\bar{I})$ and so there is an $\alpha \in F(I)$ such that $\alpha \notin P$. On the other hand $F(I) \subseteq F(R)$ so $\alpha \in F(R)$ and $\alpha \notin P \cap Z(R)$; therefore $P \cap Z(R) \in \tilde{U}_R$. But $P \cap Z(R) \notin V(I \cap Z(R))$. Finally we have to show that the stalk of the sheaf is the algebra R_P, i.e., we have to show that if $P \in \operatorname{Spec}_n(R)$ and $\mathbf{p} = P \cap Z(R)$ then $R_P \simeq R \otimes_{Z(R)} Z(R)_{\mathbf{p}}$. Now this comes from the universal property of the two localizations, as we notice that $R \otimes_{Z(R)} Z(R)_{\mathbf{p}}$ is a local algebra containing R and its maximal ideal contracts in R to the prime ideal P.

§3   Krull Dimension Revisited

We can now complete a result that we announced in (Chapter V, §4).

Theorem 3.1   If $R = F\{a_1, \ldots, a_k\}$ is a finitely generated prime *PI* algebra, Q its ring of quotients, and Z the center of Q, then $\dim R = \operatorname{Tr} \deg Z/F$.

Proof We already proved that dim $R \leqslant$ Tr deg Z/F; we claim now that in fact, if $\dim_Z Q = n^2$ then $\mathrm{Spec}_n(R) \subseteq \mathrm{Spec}(R)$ has Krull dimension exactly equal to Tr deg Z/F. In fact we have proved that $\mathrm{Spec}_n(R)$ is homeomorphic to U_R; on the other hand U_R is homeomorphic to a certain open subsheme of $\mathrm{Spec}(Z(R))$, with $Z(R) = F + F(R)$ and $F(R)$ the Formanek center of R. This open set is a union of spaces $\mathrm{Spec}(Z(R)[1/\alpha])$ where $Z(R)[1/\alpha]$ is finitely generated over F and

$$\mathrm{Tr\ deg}\ Z(R)[1/\alpha]/F = \mathrm{Tr\ deg}\ Z/F.$$

As a corollary we complete some results in Chapter V, §4.

Theorem 3.2 Let $R = F\{a_1, \ldots, a_k\}$ be an algebra of degree n over a field F. Let N be the ideal of R generated by the polynomial identities of $n \times n$ matrices.

(1) N is a nil ideal.
(2) Krull dim $R \leqslant kn^2 - (n^2 - 1)$ and it is equal to this number if and only if R/N is isomorphic to the algebra of k generic $n \times n$ matrices.

Proof (1) Since R is of degree n, then R/P with P any prime ideal is of degree $\leqslant n$ and so satisfies the identities of $n \times n$ matrices. Therefore, $N \subset \bigcap P$ and N is a nil ideal.
(2) Clearly the Krull dimension of R and R/N are equal, so we consider $\bar{R} = R/N = F\{\bar{a}_1, \ldots, \bar{a}_k\}$. Since $\bar{R}$ satisfies the identities of $n \times n$ matrices we have a surjection $F\{\xi_1, \ldots, \xi_k\} \to F\{a_1, \ldots, a_k\}$ where $\xi_1, \ldots, \xi_k$ are generic matrices.
Now we have computed the transcendence degree of the center of the ring of quotients of $F\{\xi_1, \ldots, \xi_k\}$; this number is $kn^2 - (n^2 - 1)$ (Chapter IV, **5.3**). Therefore the theorem is proved.

More generally we can prove the following result.

Proposition 3.3 Let $R \subseteq S = R\{a_1, \ldots, a_k\}$ be a finitely generated extension of *PI* rings. Furthermore, let Q be a prime ideal of S, with $P = Q \cap R$. If S is of degree n we have

$$rkQ \leqslant rkP \cdot (kn^2 - (n^2 - 1)).$$

Proof   Let $Q \supset Q_1 \supset Q_2 \supset \ldots \supset Q_n$ be a descending chain of prime ideals of S. The Q_i's contract to prime ideals P_i of R so the theorem is proved if we can show the validity of the following lemma.

Lemma 3.4   Let $R \subseteq S = R\{a_1, \ldots, a_k\}$ be a finitely generated extension of *PI* rings. Let S be of degree n and P a prime ideal of R; then any chain $Q_0 \supset Q_1 \supset Q_2 \supset \ldots \supset Q_m$ of prime ideals of S with $Q_i \cap R = P$ is of length at most $kn^2 - (n^2 - 1)$.

Proof   Let us consider such a chain; we pass to $\bar{S} = S/Q_m$. This is a prime algebra containing $\bar{R} = R/P$, and $\bar{S} = \bar{R}\{\bar{a}_1, \ldots, \bar{a}_k\}$ is an extension. Now let K_1, K_2 be the rings of quotients of $\bar{R}$, $\bar{S}$ respectively. We have $K_1 \subseteq K_2$ in a natural way and this is an extension. If Q is any prime ideal of S with $Q \cap R = P$ then $S/Q \supseteq \bar{R}$ and the ring of quotients of S/Q contains K_1. It follows, by a simple argument, that the prime ideals of S contracting to 0 in R are in a 1-1 correspondence with the prime ideals of the ring $T = K_1\{\bar{a}_1, \ldots, \bar{a}_k\} \subseteq K_2$. Now if Z is the center of K_1 we have $T = K_1 \otimes_Z Z\{\bar{a}_1, \ldots, \bar{a}_k\}$. The lattice of ideals of T is isomorphic to the lattice of ideals of $Z\{\bar{a}_1, \ldots, \bar{a}_k\}$ via the correspondence $I \leftrightarrow K_1 \otimes I$; therefore the claim follows.

BIBLIOGRAPHY

[1] ALBERT A. A. Structure of Algebras, A. M. S. Colloquium Publ. v. XXIV.

[2] AMITSUR S. A. Nil P. I.-rings, Proc. A. M. S., **2**, 538–540 (1951).

[3] — An embedding of PI-rings, Proc. A. M. S., **3** (1952).

[4] — The problem of Kurosh-Levitzki-Jacobson, Riveon Lematematika, **5**, 41–48 (1952).

[5] — The identities of PI-rings, Proc. A. M. S., **4**, 27–34 (1953).

[6] — Application to a polynomial identity, Riveon Lematematika, **7**, 30–32 (1954).

[7] — Generic splitting fields of central simple algebras, Ann. Math. (2), **62**, 8–43 (1955).

[8] — Identities and generators of matrix rings, Bull. Res. Council Israel, S. A., **5**, 5–10 (1955).

[9] — On rings with identities, J. London Math. Soc., **30**, 470–475 (1955).

[10] — Some results on central simple algebras, Ann. Math. (2), **63**, 285–293 (1956).

[11] — Radicals of polynomial rings, Canad. J. Math., **8**, 355–361 (1956).

[12] — A generalization of Hilbert's Nullstellensatz, Proc. A. M. S., **8**, 643–656 (1957).

[13] — Rings with a pivotal monomial, Proc. A. M. S., **9**, 635–642 (1958).

[14] — Finite dimensional central division algebras, Proc. A. M. S., **11**, 28–31 (1960).

[15] — Groups with representations of bounded degree II, Ill. J. Math., **5**, 198–205 (1961).

[16] — Nil semigroups of rings with a polynomial identity, Nagoya Math. J., **27**, 103–111 (1966).

[17] — Prime rings having polynomial identities with arbitrary coefficients, Proc. A. M. S. (3), **17**, 470–486 (1967).

[18] — A noncommutative Hilbert basis theorem and subrings of matrices, Trans. Amer. Math. Soc., **149**, 133–142 (1970).

[19] Embeddings in matrix rings, Pacific J. Math, **36** 21–27 (1971).

[20] AMITSUR S. A., LEVITZKI J. Minimal identities for algebras, Proc. A. M. S., **1**, 449–463 (1950).

[21] — Remarks on minimal identities for algebras, Proc. A. M. S., **2**, 320–327 (1951).

[22] AMITSUR S. A., PROCESI C. Jacobson rings and Hilbert algebras with polynomial identities, Annali Mat. Pura Applicata, **71**, 61–72 (1960).

[23] ARTIN M. On Azumaya algebras and finite dimensional representations of rings, J. Algebra, 11, 532–563 (1969).

[24] AUSLANDER M., GOLDMAN O. Maximal orders, Trans. A. M. S., 97, 1–24 (1960).

[25] — The Brauer group of a commutative ring, Trans. A. M. S., 97, 367–409 (1960).

[26] AZUMAYA G. On maximally central algebras, Nagoya Math. J., v. 2, 119–150 (1951).

[27] BOURBAKI N. Algébre, Chap. 4, 5, Hermann, Paris (1959).

[28] — Algébre Commutative, Chap. I, II, Hermann, Paris (1961).

[29] CHACRON M. On a theorem of Procesi, J. Algebra, v. 15, 225–228 (1970).

[30] CHATTERS Localization in PI-rings, J. London Math. Soc., 2, 763–768 (1970).

[31] COHN P. M. Universal Algebre, Harper (1965).

[32] CURTIS C. W. Noncommutative extensions of Hilbert rings, Proc. A. M. S., 4, 945–955 (1953).

[33] DIEUDONNÉ J., CARRELL J. B. Invariant theory, old and new, Adv. Math., 4, 1–80 (1970).

[34] DRAZIN M. P. A generalization of polynomial identities in rings, Proc. A. M. S., 8, 352–361 (1957).

[35] FORMANEK E. Central Polynomials for matrix rings, J. Algebra, 23, 129–133 (1972).

[36] GOLDMAN O. Hilbert rings and the Hilbert Nullstellensatz, Math. Zeit., v. 54, 136–140 (1951).

[37] GOLOD E. S. On nil algebras and finitely approximable groups, Izv. Adad. Nank SSSR, Ser. Mat. 28, 273–276 (1964).

[38] GOLOD E. S., SHAFAREVITCH I. R. On towers of class field, Izv. Akad. Nank SSSR, Ser. Mat. 28, 261–272 (1964).

[39] GRACE J. H., YOUNG A. The algebra of invariants, Cambridge (1903).

[40] GROTHENDIECK A. Le groupe de Brauer I, II, III., Dix Exposés sur la Cohomologie des Schémas, 46–188, North Holland, Amsterdam, Massau, Paris (1968).

[41] GROTHENDIECK A., DIEUDONNÉ J. Element de Geometrie Algebrique,   ·

[42] HARRIS B. Commutators in division rings, Proc. A. M. S., 9, 628–630 (1958).

[43] HERSTEIN I. N. Topics in ring theory, Univ. of Chicago Press (1969).

[44] — Noncommutative rings, Carus Monograph, 15, (1968).

[45] HILBERT D. Über die vollen Invariantensysteme, Math. Ann., 42, 313–373 (1893).

[46] JACOBSON N. Structure theory for algebraic algebras of bounded degree, Ann. Math. (2), 46, 695–707 (1945).

[47] — Structure of rings, A. M. S. Colloquium Publ. v. XXXVII (1964).

[48] KAPLANSKY I. Rings with a polynomial identity, Bull. A. M. S., 54, 575–580 (1948).

[49] — Groups with representations of bounded degree, Canad. J. Math., 1, 105–112 (1949).

[50] — Topological representations of algebras II, Trans. A. M. S., 8, 62–75 (1950).

[51] — Fields and rings, Univ. of Chicago Press (1969).

[52] — Problems in the theory of rings revisited, Amer. Math. Monthly, 77, 445–454 (1970).

[53] KIRILLOV A. A. On certain division algebras over rational function fields, Funkcional Anal. i, Prilozen, —, 101–102 (1967).

[54] KRULL W. Jacobsonsche Ringe, Hilbertscher Nullstellensatz, Dimension theorie, Math. Zeit., v. 54, 354–387 (1951).

[55] LERON U., VAPNE A. Polynomial identities of related rings, Israel J. Math., 8, 127–137 (1970).

[56] LEVI F. W. On skewfields of given degree, Indian Math. Soc. (N. S.), **11**, 85–88 (1947).

[57] — Über der Kommutativitätsrang in einen Ringe, Math. Ann., **121**, 184–190 (1949).

[58] LEVITZKI J. On a problem of A. Kurosh, Bull. A. M. S., **52**, 1033–1035 (1946).

[59] — A theorem on polynomial identities, Proc. A. M. S., **1**, 334–341 (1950).

[60] — On the rank of commutativity, Riveon Lematematika, **6**, 1–14 (1953).

[61] — On minimal central identities, Riveon Lematematika, **8**, 41–58 (1954).

[62] MARTINDALE W. Prime rings satisfying a generalized polynomial identity, Algebra, **12**, 576–584 (1969).

[63] MATSUMURA H. Commutative algebra, W. A. Benjamin (1970).

[64] MUMFORD D. Geometric Invariant Theory, Springer, Berlin (1965).

[65] POSNER E. C. Prime rings satisfying a polynomill identity, Proc. A. M. S., **11**, 180–183 (1960).

[66] PROCESI C. The Burnside problem, J. Algebra, **4**, n. 3, 421–425 (1966).

[67] — Noncommutative affine rings, Atti Acc. Naz. Lincei, s. VIII, v. VIII, f. 6, 239–255 (1967).

[68] — Noncommutative Jacobson rings, Annali Sci. Norm. Sup. Pisa, v. XXI, f. II, 381–390 (1967).

[69] — Sugli anelli commutativi zero dimensionali con identitá polinomiale, Rend. Circolo Mat. Palermo, s. II, T. XVII, 5–11 (1968).

[70] — Sulle identitá delle algebre semplici, Rend. Circolo Mat. Palermo, s. II, T. XVII, 13–18 (1968).

[71] — Dipendenza integrale sulle algebre non commutative, Ist. Naz. Alta Mat., Symposia Matematica VIII, 295–308 (1972).

[72] On a theorem of M. Artin **22**, 309–315 (1972).

[72] On a theorem of M. Artin (To appear in the Journal of Algebra).

[73] — Sulle rappresentazioni degli anelli e loro invarianti, Ist. Naz. Alta Mat., Symposia Matematica, (To appear).

[74] PROCESI C., SMALL L. Endomorphism rings of modules over PI-algebras, Math. Zeit., **106**, 178–180 (1968).

[75] REGEV A. Existence of identities in A ± B, Israel Journal Math., **11** (2), 131–152 (1972).

[76] SMALL L. An example in PI-rings, J. Algebra, **17**, 434–436 (1971).

[77] TOMIYAMA J., TAKESAKI M. Applications of fibre bundles to a certain class of C*-algebras, Tohoku Math. J., **13**, 498–522 (1961).

[78] VASIL'EV N. B. C*-algebras with finite dimensional irreducible representations, Russian Math. Surveys, **21**, 137–155 (1966).

[79] WEIL H. The classical groups, Princeton Univ. Press (1946).

OPEN PROBLEMS

Identities of $n \times n$ matrices

a)   It is known that the minimal identity for $n \times n$ matrices is S_{2n}, what about the minimal identity in a certain fixed number of variables $K < 2n$?

b)   More generally how can one describe all identities of $n \times n$ matrices or all multilinear identities or compute the dimension of the space of homogeneous identities of a certain degree in a certain number of variables etc.

c)   Related to the previous question is the following: if $R = K\{a_1, \ldots a_K\}$ is a finitely generated graded PI-algebra let $f(m) = \dim_K R_m$, what can one say on the function $f(m)$? Is it a polynomial function in some sense? One can see in some cases that $f(m)$ has the behavior at infinity of a polynomial function whose degree is related to the Krull dimension of the algebra R.

d)   Consider the T-ideal of identities of $n \times n$ matrices in a certain number of variables, is it finitely generated as an ideal or as a T-ideal?

e)   Consider the rings $Z\{\xi_1, \ldots, \xi_m\}$, $Z/(p)\{\xi_1, \ldots, \xi_m\}$ of generic matrices over Z and $Z/(p)$ respectively is the kernel of the natural map $Z\{\xi_1, \ldots, \xi_m\} \to Z/(p)\{\xi_1, \ldots, \xi_m\}$ generated by p?

Representation theory

a) If $R = K\{a_1, \ldots, a_k\}$ is a finitely generated algebra, the space of irreducible m dimensional representations has dimension $\psi(m)$, what can one say about $\psi(m)$?

b) One can study very well the equivalence of irreducible representations of a fixed dimension n. In particular such representation from an open set over which the projective linear group operates giving rise to a principal bundle. What can one say of the complement? In particular one can study completely reducible representations and divide them in groups according to the isomorphism type of the algebras generated by the image of the representation and then study the action of the projective linear groups on such sets.

Finitely generated algebras

Given a finitely generated PI-algebra $R = F\{a_1, \ldots, a_K\}$ over a field F.

a) Is the nil radical of R nilpotent?

b) If R is prime is it true that all maximal chains of prime ideals have the same length?

c)* If R is integral over its center Λ is Λ finitely generated over F?

---

*Question c) is now proved as a consequence of Theorem 3, Addendum to Chapter VI.

LIST OF BASIC NOTATIONS

$(A)_n$	total ring of $n \times n$ matrices over a ring A
$\mathcal{M}(A, B)$	set of homomorphisms between two algebras
$\Lambda\{x_s\}$	free algebra over Λ in the variables x_s
$\Lambda\{\xi_s\}$	ring of generic matrices
$J(R)$	Jacobson radical of a ring R
$L(R)$	lower nil radical of a ring R
$A*B$	free product of two rings
$\mathcal{S}_n$	symmetric group on n elements
$\mathcal{P}(M, N)$	polynomial maps between two modules
S^R	centralizer of R in S
$r(I)$	right annihilator of I
$l(I)$	left annihilator of I
$Tr(a)$	reduced trace of an element in a central simple algebra
$tr \deg F/G$	transcendence degree of a field extension
∂a	degree of an element in a graded algebra
$rk(P)$	rank of a prime ideal

INDEX

Other books
of interest
to you...

Because of your interest in our books, we have included the following catalog of books for your convenience.

Any of these books are available on an approval basis. This section has been reprinted in full from our **mathematics/ statistics** catalog.

If you wish to receive a complete catalog of MDI books, journals and encyclopedias, please write to us and we will be happy to send you one.

MARCEL DEKKER, INC.
95 Madison Avenue, New York, N.Y. 10016

mathematics
and statistics

BABAKHANIAN *Cohomological Methods in Group Theory*

(Pure and Applied Mathematics Series, Volume 11)

by ARARAT BABAKHANIAN, *University of Illinois, Urbana*

254 pages, illustrated. 1972.

Provides the reader with the basic tools in cohomology of groups and illustrates their use in obtaining group theoretic results, and may be used as a text for graduate students who have taken one course in algebra. Also recommended to group theorists interested in employing cohomological methods in their research.

CONTENTS: Elements of homological algebra • Cohomology of finite groups • Computations • Lower central series and dimension subgroups • Relations with subgroups and quotient groups • Finite p-nilpotent groups • Cup products • Spectral sequences • Descending central series • Galois groups.

BARROS-NETO *An Introduction to the Theory of Distributions*

(Pure and Applied Mathematics Series, Volume 14)

by JOSE BARROS-NETO, *Rutgers—The State University, New Brunswick, New Jersey*

232 pages, illustrated. 1973

Provides an introduction to the theory of distributions as it is derived from the general theory of topological vector spaces. Specially designed for the student who has taken courses on advanced calculus, linear algebra, and general topology, and who is familiar with Lebesgue integration theory and Banach and Hilbert spaces, including the Hahn-Banach theorem. Problems at the end of each chapter serve to help the reader check his comprehension of the subject matter.

CONTENTS: Locally convex spaces • Distributions • Convolutions • Tempered distributions and their Fourier transforms • Sobolev spaces • On some spaces of distributions • Applications.

BOOTHBY and WEISS *Symmetric Spaces: Short Courses Presented at Washington University*

(Pure and Applied Mathematics Series, Volume 8)

edited by WILLIAM M. BOOTHBY and GUIDO L. WEISS, *Washington University, St. Louis, Missouri*

504 pages, illustrated. 1972

Contains material presented in a series of short courses on the geometry and harmonic analysis of symmetric spaces given by well known authorities in the field. This text is of great interest to graduate students and professional mathematicians working in the areas of harmonic analysis, differential geometry, topological groups, and symmetric spaces.

CONTENTS: Minimal immersions of symmetric spaces into spheres, *N. R. Wallach.* Spherical functions on semisimple Lie groups, *R. Gangolli.* Spectra of discrete uniform subgroups of semisimple Lie groups, *R. Gangolli.* Fourier decompositions of certain representations, *K. I. Gross and R. A. Kunze.* Conical distributions and group representations, *S. Helgason.* Geometric ideas in Lie group harmonic analysis theory, *R. Hermann.* Bounded symmetric domains and holomorphic discrete series, *A. W. Knapp.* Schwarz lemma, *S. Kobayashi.* On tube domains, *Y. Matsushima.* Fine structure of Hermitian symmetric spaces, *J. A. Wolf.* Boundaries of Riemannian symmetric spaces, *H. Furstenberg.* Harmonic functions on symmetric spaces, *A. Koranyi.* Fatou's theorem for symmetric spaces, *N. J. Weiss.* New and old results in invariant theory with applications to arithmetic groups, *S. J. Rallis.* Topics on totally discontinuous groups, *H.-C. Wang.*

BURCKEL *Characterizations of C(X) Among Its Subalgebras*

(Lecture Notes in Pure and Applied Mathematics Series, Volume 6)

by R. B. BURCKEL, *Kansas State University, Manhattan*

176 pages, illustrated. 1972

(continued)

BURCKEL *(continued)*

Gives a detailed account of some recent results about subalgebras of C(X). Readers should have completed a course in real-variables and should have some acquaint-ance with functional analysis and complex-variables. Of interest to advanced under-graduates, graduate students, and profes-sional mathematicians alike.

CONTENTS: Bishop's Stone-Weierstrass theo-rem • Restriction algebras determining C(X) • Wermer's theorem on algebras with multi-plicatively closed real part • The work of Alain Bernard • The theorems of Gorin and Čirka • Bounded approximate normality, the work of Badé and Curtis • Katznelson's bounded idem-potent theorem • Characterization of C(X) by functions which operate.

CHAVEL Riemannian Symmetric Spaces of Rank One

(Lecture Notes in Pure and Applied Mathematics Series, Volume 5)

by Isaac Chavel, *City College of the City University of New York*

96 pages, illustrated. 1972

Provides an introduction to results on Riemannian symmetric spaces of rank one. A unified treatment aimed at the student familiar with the fundamentals of Rieman-nian geometry.

CONTENTS: Variational theory and com-parison theorems • ¼-Pinched manifolds • Riemannian homogeneous spaces • Riemannian symmetric spaces of rank one.

DIEUDONNÉ Introduction to the Theory of Formal Groups

(Pure and Applied Mathematics Series, Volume 20)

by J. Dieudonné, *University of Nice, France*

248 pages, illustrated. 1973

CONTENTS: Definition of formal groups • Infinitesimal formal groups • Infinitesimal com-mutative groups • Representable reduced infinitesimal groups.

DORNHOFF Group Representation Theory

In 2 Parts

(Pure and Applied Mathematics Series, Volume 7)

by Larry Dornhoff, *University of Illi-nois, Urbana*

Part A   Ordinary Representation Theory
264 pages, illustrated. 1971

Part B   Modular Representation Theory
266 pages, illustrated. 1972

Provides a readable account of several major applications of representation theory to the structure of finite groups. Serves well as a textbook for a graduate course in rep-resentation theory and is useful for indi-vidual study by graduate students and mathe-maticians wishing to familiarize themselves with the subject.

CONTENTS:

Part A: Introduction • Theory of semisimple rings • Semisimple group algebras • Splitting fields and absolutely irreducible modules • Characters • Burnside's $p^a q^b$ theorem • Multi-plicities, generalized characters, character tables • Representations of Abelian groups • Induced characters • Representations of direct products • Permutation groups • T. I. sets and exceptional characters • Frobenius groups • Clifford's theorem • M-groups • Brauer's char-acterization of characters • Brauer's theorem on splitting fields • Normal p-complements and the transfer • Generalized quaternion Sylow 2-subgroups • A theorem of Tate • Mackey de-composition • Itô's theorem on character de-grees • Algebraically conjugate characters • The Schur index • Projective representations • The finite two-dimensional linear groups • Special conjugacy classes • A characterization via centralizers of involutions • Primitive com-plex linear groups • Jordan's theorem à la Blichfeldt • Extra-special p-groups • Normal p-subgroups of primitive linear groups • The Frobenius-Schur count of involutions • Primi-tive solvable linear groups • Simplicity of $PSL(n,F)$ and $PSp(2m,F)$ • Jordan's theorem for solvable groups • Itô's theorem on char-acters of solvable groups • Characters of $SL(2, p^n)$.

Part B: Indecomposable modules and chain conditions • The radical of a ring • Idempotents • Completeness • Unique decomposition theo-rems • Lifting idempotents • Principal inde-composable modules • Cartan invariants • The number of irreducible modules • Decomposition numbers • Finite extensions of complete local domains • Existence of a suitable ring: p-adic integers • Relatively projective RG-modules • Green's theorem • Vertices and sources • De-fect groups • Central characters • The Brauer homomorphism • The Brauer correspondence • Brauer's first main theorem • Brauer characters • Orthogonality relations • Characters in blocks • Blocks of defect zero • Higher decomposition numbers; Brauer's second main theorem • Ex-tension of the first main theorem to $DC_G(D)/D$ • The principal block • Quaternion Sylow 2-subgroups • The Z^*-theorem of Glauberman • Blocks with cyclic defect group • Brauer: Groups of degree $<(p-1)/2$ • Feit and Thomp-son: Groups of degree $<(p-1)/2$ • p-blocks of $SL(2, p)$ • p-blocks of p-solvable groups.

GILMER *Multiplicative Ideal Theory*

(Pure and Applied Mathematics Series, Volume 12)

by ROBERT GILMER, *The Florida State University, Tallahassee*

624 pages, illustrated. 1972

Presents a comprehensive and detailed account of some of the important aspects of commutative algebra. Contains over 1,000 exercises and a bibliography of more than 300 entries. Directed to graduate students in mathematics and mathematicians concerned with commutative algebra.

CONTENTS: Basic concepts • Integral dependence • Valuation theory • Prüfer domains • Polynomial rings • Domains of classical ideal theory.

GRAY and SCHUCANY *The Generalized Jackknife Statistic*

(Statistics Series, Volume 1)

by H. L. GRAY, *Texas Tech University, Lubbock,* and W. R. SCHUCANY, *Southern Methodist University, Dallas, Texas*

320 pages, illustrated. 1972

Explores the theory surrounding the jackknife method and gives examples of its applications. Written from a theoretical and practical point of view and serves both the research statistician and the practitioner. Directed to statisticians, economists, professors in business colleges, psychologists, biometricians, and mathematicians.

CONTENTS: Reduction of bias by jackknifing • Applications to biased estimators • Asymptotic distributions • Jackknifing stochastic processes • The $J_\infty^{(2)}$-estimator.

HERMANN *Geometry, Physics, and Systems*

(Pure and Applied Mathematics Series, Volume 18)

by ROBERT HERMANN, *Rutgers—The State University, New Brunswick, New Jersey*

Part I in preparation. 1973

Deals with the interaction between geometry and physics and related disciplines, such as optimal control and systems theory, and continuum mechanics. Emphasizes the role of the theory of linear differential operators on manifolds, Ehresmann's theory of jets, and the geometric contributions of Sophus Lie and Elei Cartan. Requires only basic knowledge of differential and integral calculus on manifolds.

CONTENTS:
Part I: Algebraic concepts of manifold theory of differential operators • General differential-geometric form of the equations of motions of particle mechanics • Contact transformations and differential equations • Symmetries of variational problems • The theory of exterior differential systems • The mathematics of thermodynamics.

HIRZEBRUCH, NEUMANN, and KOH *Differentiable Manifolds and Quadratic Forms*

(Lecture Notes in Pure and Applied Mathematics Series, Volume 4)

by FRIEDRICH E. HIRZEBRUCH and WALTER D. NEUMANN, *University of Bonn, Germany,* and SEBASTIAN S. KOH, *West Chester State College, Pennsylvania*

128 pages, illustrated. 1971

Based on Sebastian Koh's notes on Friedrich Hirzebruch's lecture series given at Brandeis University and at the University of California at Berkeley. Explores invariants of quadratic forms and the calculation of some Grothendieck rings of quadratic forms. Also deals with manifolds and invariants defined by their quadratic forms.

CONTENTS: Quadratic forms • The Grothendieck ring • Certain arithmetical properties of quadratic forms • Integral unimodular quadratic forms • Quadratic forms over $Z(p)$; the genus of integral forms • The quadratic form of a 4k-dimensional manifold • An application of Rohlins theorem; μ-invarients • Plumbing • Complex manifolds of complex dimension 2 • A theorem of Kervaire and Milnor.

JACOBSON *Exceptional Lie Algebras*

(Lecture Notes in Pure and Applied Mathematics Series, Volume 1)

by NATHAN JACOBSON, *Yale University, New Haven, Connecticut*

136 pages, illustrated. 1971

Presents a set of models for the exceptional Lie algebras over algebraically closed fields of characteristic 0 and over the field of real numbers. An excellent tool for the mathematical public in general — especially those interested in the classification of Lie algebras or groups — and for theoretical physicists.

CONTENTS: Jordan algebras of symmetric bilinear forms • Cayley algebras • Exceptional Jordan algebras • Automorphisms of $\mathscr{J}_4$'s • f_4 and $\mathscr{E}^6$ • Lie algebras of type E_6 • Some Exceptional Lie algebras of type D_4 • Roots of applications of Galois cohomology • Lie alge-

(continued)

JACOBSON *(continued)*

bras of type E_7 • Tits' second construction • Calculation of the Killing forms • Models of the real forms.

KOBAYASHI Hyperbolic Manifolds and Holomorphic Mappings

(Pure and Applied Mathematics Series, Volume 2)

by SHOSHICHI KOBAYASHI, *University of California, Berkeley*

160 pages, illustrated. 1970

Presents a coherent account of intrinsic pseudo distances on complex manifolds and their applications to holomorphic mappings. Of value to graduate students and research mathematicians interested in differential geometry, complex analysis, and related fields.

CONTENTS: The Schwarz lemma and its generalizations • Volume elements and the Schwarz lemma • Distance and the Schwarz lemma • Invariant distances on complex manifolds • Holomorphic mappings into hyperbolic manifolds • The big Picard theorem and extension of holomorphic mappings • Generalization to complex spaces • Hyperbolic manifolds and minimal models • Miscellany.

KSHIRSAGAR Multivariate Analysis

(Statistics Series, Volume 2)

by ANANT M. KSHIRSAGAR, *Texas A&M University, College Station*

552 pages, illustrated. 1972

Deals with the advanced theory of multivariate analysis and describes how multivariate techniques may be applied to practical problems in anthropometry, econometrics, psychometry, agriculture, and biometry.

Excellent as a one- or two-semester textbook for graduate students interested in multivariate analysis and also of importance to students and research scientists in statistics, and those dealing with statistical problems in anthropology, biology, and the agricultural and medical sciences.

CONTENTS: Regression and correlation among several variables • Multivariate normal distribution • The Wishart distribution • Distributions associated with regression • Hotelling's T^2 and its applications • Discriminant analysis • Canonical variables and canonical correlations • Wilks' Λ criterion • Multivariate analysis of variance and discrimination in the case of several groups • Likelihood ratio tests • Principal components.

LARSEN Functional Analysis

(Pure and Applied Mathematics Series, Volume 15)

by RONALD LARSEN, *Wesleyan University, Middletown, Connecticut*

512 pages, illustrated. 1973

A text which provides the basic theories, results, and techniques of functional analysis, as well as a sampling of its various applications to problems in functional analysis and other areas of mathematics. Includes problem sets for each chapter. For graduate and advanced undergraduate students who are acquainted with point set topology, linear algebra, complex analysis, and measure and integration theory.

CONTENTS: Semi–normed and normed linear spaces • Topological linear spaces • Linear transformations and linear functionals • The Hahn–Banach theorem: Analytic form • The Hahn–Banach theorem: Geometric form • The uniform boundedness theorem • The open mapping theorem and the closed graph theorem • Reflexity • Weak topologies • The Krein–Smulian theorem and the Eberlein–Smulian theorem • The Krein–Mil'man theorem • Fixed point theorem • Hilbert space.

LINDAHL and POULSEN Thin Sets in Harmonic Analysis

(Lecture Notes in Pure and Applied Mathematics, Volume 2)

edited by L.-Å. LINDAHL, *Uppsala University, Sweden*, and F. POULSEN, *University of Aarhus, Denmark*

200 pages, illustrated. 1971

Based on a seminar on thin sets presented at the Institute Mittag-Leffler, Djursholm, Sweden, during the year 1969-70. Valuable as an introduction to the field and as a text on which to base seminars.

CONTENTS: Dirichlet, Kronecker, and Helson sets, *L.-Å. Lindahl*. Bernard's theorem, *T. W. Körner*. Harmonic synthesis of Kronecker sets, *U. Tewari*. Continuous curves whose graphs are Helson sets, *T. Hedberg*. Sidon sets, *J.-E. Björk*. Combinatorial methods and Sidon sets, *F. Poulsen*. The characters of $S(K)$ and the union problem for a Kronecker set and a point, *N. Th. Varopoulos*. Bochner's theorem for $S(K)$, *N. Th. Varopoulos*. The group $S^*(X_{,\mu})$, *N. Th. Varopoulos*. The union problem for Helson sets, *N. Th. Varopoulos*. The embedding of $A(E)$ into $\tilde{A}(E)$, *N. Th. Varopoulos*. Unbounded synthesis, *N. Th. Varopoulos*. Relations between Dirichlet, Kronecker, and Helson sets, *T. W. Körner*. Ultra-thin symmetric sets and harmonic analysis, *Y. Meyer*. Pisot numbers and the problem of synthesis, *Y. Meyer*.

MATSUSHIMA *Differentiable Manifolds*

(Pure and Applied Mathematics Series, Volume 9)

by YOZO MATSUSHIMA, *University of Notre Dame, Indiana*

translated by EDWARD KOBAYASHI

320 pages, illustrated. 1972

An introduction to the theories of differentiable manifolds and Lie groups. Designed as a continuation of advanced calculus.

CONTENTS: Introduction • Differentiable manifolds • Differential forms and tensor fields • Lie groups and homogeneous spaces • Integration of differential forms and their applications.

MENDEL *Discrete Techniques of Parameter Estimation: The Equation Error Formulation*

(Control Theory Series, Volume 1)

by JERRY M. MENDEL, *McDonnell Douglas Astronautics Company, Huntington Beach, California*

408 pages, illustrated. 1973

Provides a solid analytical foundation for the estimation techniques of generalized least-squares, unbiased minimum-variance, deterministic-gradient, and stochastic-gradient. Presents these techniques from a unified point of view made possible by the use of the Equation Error Method as the basis for problem formulation. Problems accompany each chapter. Valuable as a graduate text in a course on techniques of parameter estimation and also of interest to statisticians, econometricians, and aerospace and control engineers.

CONTENTS: Equation error formulation of parameter estimation problems • Least-squares parameter estimation • Minimum-variance parameter estimation • Deterministic-gradient parameter estimation • Stochastic-gradient parameter estimation • Estimation of time-varying parameters.

NACHBIN *Convolutions and Holomorphic Mappings*

(Pure and Applied Mathematics Series)

by LEOPOLDO NACHBIN, *University of Rochester, New York*

in preparation. 1973

The first part of the text describes the work by C. P. Gupta and the author on nuclearly entire functions on a complex Banach space and the approximation of

solutions of a convolutional equation by finite sums of exponential-polynomials. The second part surveys work by S. B. Chae, S. Dineen and the author on topological properties of locally convex spaces of holomorphic mappings between Banach spaces.

CONTENTS (partial): Convolutions on nuclearly entire functions • Spaces of holomorphic mappings.

NARICI, BECKENSTEIN, and BACHMAN *Functional Analysis and Valuation Theory*

(Pure and Applied Mathematics Series, Volume 5)

by LAWRENCE NARICI, *St. John's University, Jamaica, New York,* EDWARD BECKENSTEIN and GEORGE BACHMAN, *Polytechnic Institute of Brooklyn, New York*

200 pages, illustrated. 1971

"It is clearly written. . . . Exceptionally well supplied with exercises. . . . The material is well chosen and at a high scientific level."—Jack Schwartz, *Courant Institute*

CONTENTS: Topological results • Completeness • Normed linear spaces • Normed algebras.

NEY and PORT *Advances in Probability and Related Topics*

a series edited by PETER NEY, *University of Wisconsin, Madison,* and SIDNEY PORT, *University of California, Los Angeles*

Vol. 1 236 pages, illustrated. 1970

Vol. 2 264 pages, illustrated. 1970

Vol. 3 in preparation. 1973

A series consisting of research articles and critical reviews that treat current work in probability and related topics. Directed to research mathematicians.

CONTENTS:

Volume 1: Random walks and discrete subgroups of Lie groups, *H. Furstenberg.* The technique of using random measures and random sets in harmonic analysis, *J. Kahane.* Recent developments in the theory of finite Toeplitz operators, *I. Hirschman, Jr.* Matching theory, an introduction, *G. Rota and L. Harper.*
Volume 2: Ergodic properties of operators in Lebesgue space, *M. A. Akcoglu and R. V. Chacon.* The role of reproducing kernel Hilbert spaces in the study of Gaussian processes, *G. Kallianpur.* The sums of iterates of a positive operator, *D. Ornstein.* Boundary decomposition of locally-Hunt processes, *A. O. Pittenger.* Some new stochastic integrals and Stieltjes integrals, Part I. Analogues of Hardy-Littlewood classes, *L. C. Young.*

PASSMAN   *Infinite Group Rings*

(Pure and Applied Mathematics Series, Volume 6)

by DONALD S. PASSMAN, *University of Wisconsin, Madison*

160 pages, illustrated. 1971

Offers a coherent account of the basic results in this field. Of value as a text for second year graduate students and will interest all mathematicians working with infinite group rings.

CONTENTS: Linear identities • Bounded representation degree • Nil and nilpotent ideals • Idempotents and annihilators • Research problems.

PROCESI   *Rings with Polynomial Identities*

(Pure and Applied Mathematics Series, Volume 17)

by CLAUDIO PROCESI, *University of Pisa, Italy*

192 pages, illustrated. 1973

Presents a comprehensive account of the results obtained in the last twenty years on the theory of rings with polynomial identities. Directed to graduate students and research mathematicians.

CONTENTS: Polynomial identities in algebras • Structure theorems • The identities of matrix algebras • Representations and their invariants • Finitely generated algebras and extensions • Finiteness theorems • Intrinsic characterization of Azumaya algebras • The center of a PI-ring.

PSHENICHNYI   *Necessary Conditions for an Extremum*

(Pure and Applied Mathematics Series, Volume 4)

by BORIS N. PSHENICHNYI, *Institute of Cybernetics, Kiev, U.S.S.R.*

translated by KAROL MAKOWSKI

translation edited by LUCIEN W. NEUSTADT

248 pages, illustrated. 1971

Deals with the basic principles of the modern theory of necessary conditions for an extremum. Logically develops the needed mathematical apparatus starting with very basic concepts, so that no advanced mathematical knowledge is required of the reader.

SATAKE   *Classification Theory of Semi-Simple Algebraic Groups*

(Lecture Notes in Pure and Applied Mathematics Series, Volume 3)

by ICHIRO SATAKE, *University of California, Berkeley*

160 pages, illustrated. 1971

Explains the general principle of classification theory of semi-simple algebraic groups. Includes an appendix written by M. Sugiura of the University of Tokyo simplifying previous methods of classifying real semi-simple algebraic groups. Directed toward graduate students working in algebraic groups and related areas.

CONTENTS: Preliminaries on Algebraic Groups • General Principles of Classification.

SATAKE   *Linear Algebra*

(Pure and Applied Mathematics Series)

by ICHIRO SATAKE, *University of California, Berkeley*

in preparation. 1973

Presents a self-contained account of the basic concepts of linear algebra. Employs modern algebraic methods and emphasizes their usefulness in other branches of mathematics. An excellent textbook for undergraduate students, and a useful guide for self-study.

VAISMAN   *Cohomology and Differential Forms*

(Pure and Applied Mathematics Series, Volume 21)

by IZU VAISMAN, *Semiarul Matematic Universitate, Iasu, Romania*

304 pages, illustrated. 1973

Studies cohomology spaces in three important situations: the sheaf of germs of locally constant functions on a differentiable manifold; the sheaf of germs of differentiable functions which are constant on the leaves of a foliation; and the sheaf of germs of holomorphic functions on a complex manifold. Develops the necessary subjects of homological algebra, algebraic topology, and differential geometry, and includes problems and applications. Especially written for graduate students and research workers in the various domains of differential geometry and global analysis. Also useful to any mathematician interested in categories, homological algebra, and algebraic topology.

CONTENTS: Categories and functions • Sheaves and cohomology • Fiber and vector bundles • Differential geometry • Cohomology classes and differential forms.

VLADIMIROV *Equations of Mathematical Physics*

(Pure and Applied Mathematics Series, Volume 3)

by VASILIY S. VLADIMIROV, *Steklov Institute of Mathematics, Moscow, U.S.S.R.*

translated by AUDREY LITTLEWOOD

translation edited by ALAN JEFFREY

428 pages, illustrated. 1971

Examines classical boundary value problems for differential equations of mathematical physics, using the concept of the generalized solution instead of the traditional means of presentation. Devotes a special chapter to the theory of generalized functions and may be used as a graduate text.

CONTENTS: Formulation of boundary value problems in mathematical physics • Generalized functions • Fundamental solutions and the Cauchy problem • Integral equations • Boundary value problems for elliptic equations • The mixed problem.

WALLACH *Harmonic Analysis on Homogeneous Spaces*

(Pure and Applied Mathematics Series, Volume 19)

by NOLAN WALLACH, *Rutgers—The State University, New Brunswick, New Jersey*

352 pages, illustrated. 1973

CONTENTS: Vector bundles • Elementary representation theory • Basic structure theory of compact Lie groups and semi-simple Lie algebras • The topology and representation theory of compact Lie groups • Harmonic analysis on a homogeneous vector bundle • Holomorphic vector bundles over flag manifolds • Analysis on semi-simple Lie groups • Representations of semi-simple Lie groups.

WARD *Topology: An Outline for a First Course*

(Pure and Applied Mathematics Series, Volume 10)

by LEWIS E. WARD, JR., *University of Oregon, Eugene*

128 pages, illustrated. 1972

A textbook designed for a one-year introductory course in topology. Uses the Socratic method in which students provide proofs of theorems and solutions to problems without the assistance of other ma-

terials. Emphasizes the development of deductive reasoning and includes numerous examples as an additional teaching aid.

CONTENTS: Prerequisites • Topological spaces • Connected sets • Compact sets • Separation axioms • Mappings • Product spaces • Countability axioms • Complete spaces and perfect sets • Inverse limits • Quotient spaces • Nets and compactness • Embedding and metrization • Locally compact spaces • Continua • Cutpoints and arcs • Indecomposable continua • Locally connected spaces • Arcs and mapping theorems • Partially ordered spaces • The Brouwer fixed-point theorem • Homotopy • The fundamental group.

YANO *Integral Formulas in Riemannian Geometry*

(Pure and Applied Mathematics Series, Volume 1)

by KENTARO YANO, *Tokyo Institute of Technology, Japan*

168 pages, illustrated. 1970

CONTENTS: Fundamental concepts and formulas in Riemannian geometry • Harmonic 1-forms and Killing vector fields • Riemannian manifolds admitting an infinitesimal conformal transformation • Harmonic forms and Killing tensor fields • Hypersurfaces of Riemannian manifolds • Closed hypersurfaces of a Riemannian manifold with constant mean curvature • Harmonic 1-forms and Killing vector fields in Riemannian manifolds with boundary • Harmonic forms and Killing tensor fields in Riemannian manifolds with boundary.

YANO and ISHIHARA *Differential Geometry of Tangent and Cotangent Bundles*

(Pure and Applied Mathematics Series, Volume 16)

by KENTARO YANO and S. ISHIHARA, *Tokyo Institute of Technology, Japan*

448 pages, illustrated. 1973

A summation of the results on tangent and cotangent bundles known up to the present time. Not only gives a clear insight to classical results, but also provides many new problems in the study of modern differential geometry, encouraging further investigation of this new branch of an old discipline.

CONTENTS: Vertical and complete lifts from a manifold to its tangent bundle • Horizontal lifts from a manifold to its tangent bundle • Cross-sections in the tangent bundle • Tangent bundle of Riemannian manifolds • Prolongations of G-structures to tangent bundles • Non-linear connections in tangent bundles •

(continued)

YANO and ISHIHARA *(continued)*

Vertical and complete lifts from a manifold to its cotangent bundle • Horizontal lifts from a manifold to its cotangent bundle • Tensor fields and connections on a cross-section in the cotangent bundle • Prolongations of tensor fields and connections to the tangent bundle of order 2 • Prolongations of tensor fields, connections, and G–structures to the tangent bundle of higher order.

YEH *Stochastic Processes and the Wiener Integral*

(Pure and Applied Mathematics Series, Volume 13)

by JAMES J. YEH, *University of California, Irvine*

560 pages, illustrated. 1973

A self-contained treatise of a range of closely related topics on the space of sample functions of a stochastic process. Based on courses taught by the author at the University of California at Irvine. Of particular value as a textbook for students with knowledge of real analysis and some measure theoretic probability theory.

CONTENTS: Stochastic processes • Martingales • Additive processes • Gaussia processes • Stochastic integrals • Gaussian measures in function spaces • Wiener measure and Wiener integral • Transformations of Wiener integrals.

OTHER BOOKS OF INTEREST

EL BAZ and CASTEL *Graphical Methods of Spin Algebras in Atomic, Nuclear, and Particle Physics*

by EDGARD EL BAZ, *Claude Bernard University of Lyon, France*, and BORIS CASTEL, *Queen's University, Kingston, Canada*

444 pages, illustrated. 1972

HEINMETS *Concepts and Models of Biomathematics: Simulation Techniques and Methods*

(Biomathematics Series, Volume 1)

edited by F. HEINMETS, *U. S. Army Natick Laboratories, Natick, Massachusetts.*

302 pages, illustrated. 1969

JOURNALS OF INTEREST

COMMUNICATIONS IN STATISTICS

editor: DONALD B. OWEN, *Southern Methodist University, Dallas, Texas*

Communications in Statistics is devoted to presenting the formulation and discussion of problems and their solutions, whether elegant or practical. The communication of interesting applications of known methods to real problems in industry and government will be as important as those articles having a strong mathematical orientation that have an impact on the field of statistics. The journal is reproduced directly from manuscripts submitted by authors, enabling the editor to publish articles from three to five months from the time of their receipt. *Communications in Statistics* is a vehicle for the rapid dissemination of new ideas in all areas of statistics.

6 issues per volume

TRANSPORT THEORY AND STATISTICAL PHYSICS
An International Journal for Rapid Communication in Irreversible Statistical Mechanics

editor: PAUL ZWEIFEL, *Virginia Polytechnic Institute, Blacksburg*

This new journal will be devoted primarily to those areas of irreversible statistical mechanics which are generally known as transport theory and kinetic theory. It will include papers on transport of neutral particles, dynamics of simple liquids, kinetic theory of liquids and plasmas, gas dynamics, and correlation functions.

4 issues per volume

An examination copy of any journal is available upon request.